Advanced Ceramic Coatings and Materials for Extreme Environments II

Advanced Ceramic Coatings and Materials for Extreme Environments II

A Collection of Papers Presented at the 36th International Conference on Advanced Ceramics and Composites
January 22–27, 2012
Daytona Beach, Florida

Edited by
Dongming Zhu
Hua-Tay Lin
Yanchun Zhou
Taejin Hwang

Volume Editors
Michael Halbig
Sanjay Mathur

A John Wiley & Sons, Inc., Publication

Published by John Wiley & Sons, Inc., Hoboken, New Jersey.
Published simultaneously in Canada.

Library of Congress Cataloging-in-Publication Data is available.

ISBN: 978-1-118-20589-1
ISSN: 0196-6219

10 9 8 7 6 5 4 3 2 1

Contents

ADVANCED WEAR-CORROSION RESISTANT, NANO-COMPOSITE AND MULTI-FUNCTIONAL COATINGS

THERMAL PROTECTION SYSTEMS

MATERIALS FOR EXTREME ENVIRONMENTS

Preface

The 19 papers presented in this issue include papers, which provide the up-to-date summary on the development and applications of advanced ceramic coatings and UHTC materials, from the 36th International Conference on Advanced Ceramiçs and Composites held in Daytona Beach, Florida, January 27, 2012 in the following three symposia/focused sessions:

- International Symposium on Advanced Ceramic Coatings for Structural, Environmental and Functional Applications;
- International Symposium on Materials for Extreme Environments: Ultrahigh Temperature Ceramics (UHTCs) and Nanolaminated Ternary Carbides and Nitrides (MAX Phases)
- Focused Session on Next Generation Technologies for Innovative Surface Coatings

We are greatly in debt to the members of the symposium organizing committees, for their assistance in developing and organizing these vibrant and cutting-edge symposia and focused session. We also would like to express our sincere thanks to manuscript authors and reviewers, all the symposium participants and session chairs for their contributions to a successful meeting. Finally, we are also very grateful to the staffs of The American Ceramic Society for their dedicated efforts in ensuring an enjoyable as well as successful conference and the high-quality publication of the proceeding volume.

Dongming Zhu, NASA Glenn Research Center, USA
H. T. Lin, Oak Ridge National Laboratory, USA
Yanchun Zhou, Aerospace Research Institute of Materials and Processing Technology, China
Taejin Hwang, Korea Institute of Industrial Technology, Korea

Introduction

This issue of the Ceramic Engineering and Science Proceedings (CESP) is one of nine issues that has been published based on content presented during the 36th International Conference on Advanced Ceramics and Composites (ICACC), held January 22–27, 2012 in Daytona Beach, Florida. ICACC is the most prominent international meeting in the area of advanced structural, functional, and nanoscopic ceramics, composites, and other emerging ceramic materials and technologies. This prestigious conference has been organized by The American Ceramic Society's (ACerS) Engineering Ceramics Division (ECD) since 1977.

The 36th ICACC hosted more than 1,000 attendees from 38 countries and had over 780 presentations. The topics ranged from ceramic nanomaterials to structural reliability of ceramic components which demonstrated the linkage between materials science developments at the atomic level and macro level structural applications. Papers addressed material, model, and component development and investigated the interrelations between the processing, properties, and microstructure of ceramic materials.

The conference was organized into the following symposia and focused sessions:

Symposium 1	Mechanical Behavior and Performance of Ceramics and Composites
Symposium 2	Advanced Ceramic Coatings for Structural, Environmental, and Functional Applications
Symposium 3	9th International Symposium on Solid Oxide Fuel Cells (SOFC): Materials, Science, and Technology
Symposium 4	Armor Ceramics
Symposium 5	Next Generation Bioceramics

Symposium 6	International Symposium on Ceramics for Electric Energy Generation, Storage, and Distribution
Symposium 7	6th International Symposium on Nanostructured Materials and Nanocomposites: Development and Applications
Symposium 8	6th International Symposium on Advanced Processing & Manufacturing Technologies (APMT) for Structural & Multifunctional Materials and Systems
Symposium 9	Porous Ceramics: Novel Developments and Applications
Symposium 10	Thermal Management Materials and Technologies
Symposium 11	Nanomaterials for Sensing Applications: From Fundamentals to Device Integration
Symposium 12	Materials for Extreme Environments: Ultrahigh Temperature Ceramics (UHTCs) and Nanolaminated Ternary Carbides and Nitrides (MAX Phases)
Symposium 13	Advanced Ceramics and Composites for Nuclear Applications
Symposium 14	Advanced Materials and Technologies for Rechargeable Batteries
Focused Session 1	Geopolymers, Inorganic Polymers, Hybrid Organic-Inorganic Polymer Materials
Focused Session 2	Computational Design, Modeling, Simulation and Characterization of Ceramics and Composites
Focused Session 3	Next Generation Technologies for Innovative Surface Coatings
Focused Session 4	Advanced (Ceramic) Materials and Processing for Photonics and Energy
Special Session	European Union—USA Engineering Ceramics Summit
Special Session	Global Young Investigators Forum

The proceedings papers from this conference will appear in nine issues of the 2012 Ceramic Engineering & Science Proceedings (CESP); Volume 33, Issues 2–10, 2012 as listed below.

- Mechanical Properties and Performance of Engineering Ceramics and Composites VII, CESP Volume 33, Issue 2 (includes papers from Symposium 1)
- Advanced Ceramic Coatings and Materials for Extreme Environments II, CESP Volume 33, Issue 3 (includes papers from Symposia 2 and 12 and Focused Session 3)
- Advances in Solid Oxide Fuel Cells VIII, CESP Volume 33, Issue 4 (includes papers from Symposium 3)
- Advances in Ceramic Armor VIII, CESP Volume 33, Issue 5 (includes papers from Symposium 4)

- Advances in Bioceramics and Porous Ceramics V, CESP Volume 33, Issue 6 (includes papers from Symposia 5 and 9)
- Nanostructured Materials and Nanotechnology VI, CESP Volume 33, Issue 7 (includes papers from Symposium 7)
- Advanced Processing and Manufacturing Technologies for Structural and Multifunctional Materials VI, CESP Volume 33, Issue 8 (includes papers from Symposium 8)
- Ceramic Materials for Energy Applications II, CESP Volume 33, Issue 9 (includes papers from Symposia 6, 13, and 14)
- Developments in Strategic Materials and Computational Design III, CESP Volume 33, Issue 10 (includes papers from Symposium 10 and from Focused Sessions 1, 2, and 4)

The organization of the Daytona Beach meeting and the publication of these proceedings were possible thanks to the professional staff of ACerS and the tireless dedication of many ECD members. We would especially like to express our sincere thanks to the symposia organizers, session chairs, presenters and conference attendees, for their efforts and enthusiastic participation in the vibrant and cutting-edge conference.

ACerS and the ECD invite you to attend the 37th International Conference on Advanced Ceramics and Composites (http://www.ceramics.org/daytona2013) January 27 to February 1, 2013 in Daytona Beach, Florida.

MICHAEL HALBIG AND SANJAY MATHUR
Volume Editors
July 2012

Advanced Thermal and Environmental Coating Processing and Characterizations

MULTILAYERED THERMAL BARRIER COATING ARCHITECTURES FOR HIGH TEMPERATURE APPLICATIONS

Douglas E. Wolfe[1], Michael P. Schmitt[1], Dongming Zhu[2], Amarendra. K. Rai[3] and Rabi Bhattacharya[3]

[1]The Applied Research Laboratory, The Pennsylvania State University, University Park, PA 16802
[2] NASA Glenn Research Center, Cleveland, OH 44135
[3] UES, Inc, 4401 Dayton-Xenia Road, Dayton, Ohio 45432-1894

ABSTRACT

Pyrochlore oxides of the rare earth zirconates are excellent candidates for use as thermal barrier coating (TBC) materials in highly efficient turbine engine operation at elevated temperatures (>1300°C). Pyrochlore oxides have most of the relevant attributes for use at elevated temperatures such as phase stability, low sintering kinetics and low thermal conductivity. One of the issues with the pyrochlore oxides is their lower toughness compared to the currently used TBC material viz. yttria (6 - 8 wt. %) stabilized zirconia (YSZ). In this work, arguments are advanced in favor of a multilayered coating approach to enhance erosion performance by improving toughness and lowering thermal conductivity of the TBCs. Unique multilayered coating design architectures were fabricated with alternating layers of pyrochlore oxide viz $Gd_2Zr_2O_7$ and low k TBC (rare earth oxide doped t' YSZ) utilizing electron beam physical vapor deposition (EB-PVD) technique to reduce overall rare earth oxide content. Microstructure, phase, erosion resistance and thermal conductivity of the as-fabricated multilayered coatings were evaluated and compared with that of the single layered coatings.

INTRODUCTION

Thermal barrier coatings (TBCs) are used to protect metallic components from the elevated temperatures encountered during combustion in turbine engines. Insulating the engine components allows for a longer service life as well as increased operating temperatures, and therefore increased engine efficiency. TBCs are multilayered coating systems composed of bond coat, thermally grown oxide (TGO) and ceramic topcoat. The bond coat serves as an alumina scale former to protect the underlying metallic (Ni based super alloys) components from oxidation and corrosion, as well as an intermediate coefficient of thermal expansion between the ceramic topcoat and nickel based superalloy substrate. Coating adherence is crucial, therefore the intermediate coefficient of thermal expansion of the bond coat is important in terms of prevention spallation during thermal cycling. The bond coat is often a platinum modified aluminide or MCrAlX (where M= Ni, Co and X= Y, Hf or a rare earth), with the former in the β phase and the latter in the γ phase. The β-(Ni,Pt)Al coatings have excellent cyclic oxidation properties, but lack corrosion resistance and exhibit interdiffusion with the superalloy resulting in formation of martensite or γ' and subsequent 'rumpling' due to the volume change of the phase transformation [1-3]. Moreover, interdiffusion in latest alloys containing large amounts of refractory elements results in the formation of brittle phases in a secondary reaction zone [4,5]. More recently, MCrAlX coatings have been developed with a γ-γ' phase [4]. These coatings have been shown to decrease the formation of the secondary reaction zone

The topcoat is composed of a low thermal conductivity material, typically yttria stabilized zirconia (YSZ), and is responsible for the thermal insulation of the underlying components. As engine operating temperatures continue to increase, YSZ will be insufficient due to its thermal instability above ~1200 °C. Therefore, new materials must be developed that exhibited elevated temperature phase stability (>1300 °C), low sintering kinetics and low thermal conductivity. The two frontrunners for replacement of standard YSZ are the rare earth zirconates and the rare earth doped low thermal conductivity YSZs (low-k) [6].

Pyrochlore oxides of the rare earth zirconates ($RE_2Zr_2O_7$ – where RE is a rare earth) are promising substitutes for YSZ as they exhibit elevated temperature phase stability and low thermal conductivity [7,8]. Indeed, gadolinium zirconate (GZO – $Gd_2Zr_2O_7$) has been studied by several authors due to its high thermal stability ceiling, low thermal conductivity, relatively large coefficient of thermal expansion and reduced propensity for sintering [7-11]. A drawback of GZO and other rare earth zirconates is poor erosion performance of the cubic structure and the instability with alumina TGO [6,12]. The latter can be circumvented by first depositing a YSZ interface layer before GZO deposition, thus preventing diffusion of the Gd into the TGO layer.

Low-k YSZ is a stabilized zirconia which utilizes heavy rare earth elements such as Gd or Yb to create immobile defect clusters within the structure [13]. These defect clusters serve to reduce thermal conductivity, provide sintering resistance and increase thermal stability [14]. Below approximately 6 mol% total dopant, the t' structure is exhibited, while above 6mol%, the cubic structure is stable [15]. This is important, as increasing the dopant concentration is favorable in terms of reducing thermal conductivity. On the other hand, increasing the dopant concentration in low-k ZrO_2 based systems has also been shown to decrease the thermal cyclic lifetimes [15]. Also, similarly to GZO, concerns arise over the erosion performance of the cubic phase in low-k.

This study aims to develop unique coating architectures that enhance the erosion performance of GZO while decreasing the overall TBC thermal conductivity with respect to YSZ. This will be accomplished by using a multilayer approach whereby multiple layers with either non-distinct or distinct interfaces are deposited via electron beam physical vapor deposition (EB PVD). It is expected that multiple interfaces will provide higher toughness and enhanced scattering of heat carriers[16] to decrease erosion rate and thermal conductivity, respectively. Non distinct interfaces were created in a single component of the TBC through the shuttering technique (described later) where as distinct interfaces were created by depositing alternating layers of two different components. In this work, non distinct multilayers of pure GZO and pure low-k were deposited via shutter while distinct multilayers coatings were fabricated by depositing alternating layers of GZO and low k. In such multilayer, besides interface effect, addition of tougher low-k (t') may also lower erosion rate compared to pure GZO. For comparison purpose, monolayers of 7YSZ, GZO and low-k were also fabricated. Microstructure, erosion rate and thermal conductivity of as-fabricated multilayered coating designs were determined and the results are discussed.

PROCEDURE

One inch diameter platinum aluminide coated Rene N-5 'buttons' were heat treated for 30 min in air at 704 °C and grit blasted at a 45° angle using 400 μm alumina media at 30 PSI. The samples were then tack welded to 304 stainless steel strips and the strips were subsequently tack welded to a 5.08 cm diameter mandrel. The mandrel was ultrasonically cleaned for 20 minutes in solutions of acetone and methanol, with a DI water rinse and nitrogen dry after each clean. Next, the mandrel was mounted in an industrial prototype Sciaky EB-PVD unit capable of using six EB-guns (1-6) and three independent ingot feeders (A-C) as shown in Figure 1 (a). Table 1 presents a matrix of the coating compositions and the individual deposition parameters for each of those coatings will be described below.

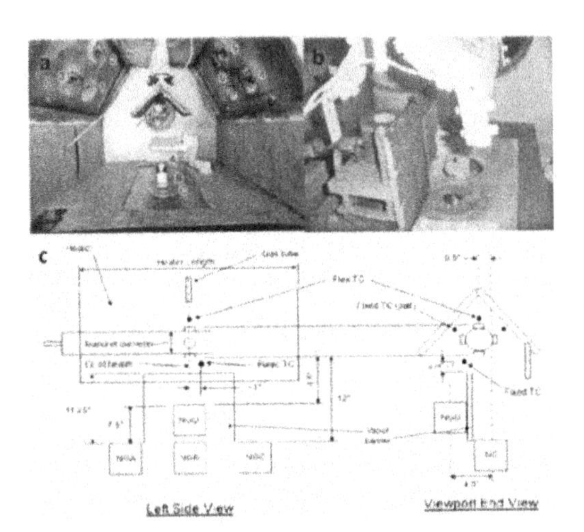

Figure 1: ARL industrial prototype EB-PVD evaporation equipment: (a) Digital image of chamber interior showing the 6 electron beam gun (labeled 1-6) and 3 continuous feed ingots (labeled A-C). Also note the 'A-Frame' heater plate, flex thermocouples, substrate mandrel and shutter in the center of the chamber. (b) Digital image of the chamber interior after setup for multi-source layering with the continuous feed source labeled as NGD. (c) A schematic illustration depicts a side and front view of the setup used for the multi-source layering depositions.

Table I: Coating composition matrix

Matrix Number	Run Number	Coating Sequence
1	S101111	YSZ, Monolayer
2	S101105	low-k YGdYbSZ, Monolayer
3	S101117	YSZ\$Gd_2Zr_2O_7$, Monolayer
4	S101119	YSZ\$Gd_2Zr_2O_7$, Multilayer (non-distinct interfaces) (40 layer)
5	S101104	low-k YGdYbSZ , Multilayer (non-distinct interfaces) (41 layers)
6	S101025	low-k YGdYbSZ\{ $Gd_2Zr_2O_7$ + / low-k YSZ} (1:1), Multilayer (distinct interfaces)
7	S101027	low-k YGdYbSZ\{low-k YSZ + / $Gd_2Zr_2O_7$} (1:4), Multilayer (distinct interfaces)

Matrix 1

The standard chamber setup was used, as shown in Figure 1 (a). A 49.5 mm diameter 7YSZ ingot (TransTech), was loaded into crucible B in Figure 1. Samples were centered above ingot B and rotated at 7 RPM and heated via electron gun #2 and gun #5 rastering on the 'A-Frame' graphite heater plate. After samples reached ~1025°C (as measured by flexible thermocouples), there was a 20 minute soak to ensure uniform substrate temperatures followed by the formation of a TGO layer by flowing oxygen at 150 sccm for another 20 minutes. Oxygen flow was then decreased to 125 sccm and a melt was established on the YSZ ingot using electron gun #6. A feed rate of 0.8 mm/min of the YSZ ingot was used during the deposition and once the desired amount of coating material was evaporated, the deposition shutter was closed and the electron guns were ramped down and turned off. The samples were allowed to cool for 10 minutes in the load lock chamber, after which the chamber was vented to atmosphere.

Matrix 2

The experimental details and procedures were identical to Matrix 1, with the exception that the ingot B material was low-k YSZ (ZrO_2 -2mol%Y_2O_3 -1mol%Gd_2O_3 -1mol%Yb_2O_3– TransTech).

Matrix 3

The initial stages of the deposition were exactly as in Matrix 1 except that ingot A was $Gd_2Zr_2O_7$ (TransTech) and ingot B was 7YSZ. Initially, gun # 6 initiated a melt on the YSZ ingot in crucible B for a total of 12.5 mm of evaporated YSZ ingot to obtain a YSZ layer approximately 25 μm thick. This layer acts as a diffusion barrier to prevent a deleterious reaction of the bond coat with the GZO. Next, the deposition shutter was closed and gun #6 was ramped down while gun #3 was brought online and a melt was started on the GZO ingot located in crucible A. The samples were then translated 160 mm directly above the GZO ingot. The deposition shutter was opened and TBC deposition was continued with a GZO feed rate of 0.8 mm/min. The shutter was then closed and the electron beam guns were ramped down and turned off and the samples were allowed to cool for 10 minutes in the load lock before the chamber was vented.

Matrix 4

The experimental details and procedures were identical to Matrix 3, with the exception that, during the GZO deposition, the depositing vapor flux was interrupted using the deposition shutter to create non-distinct interfaces. The deposition shutter was open for a period of 2.3 minutes and closed for 0.5 minutes and this was repeated for a total of 40 layers.

Matrix 5

The experimental details and procedures were identical to Matrix 1, with two exceptions. First, the ingot B material was low-k YSZ (ZrO_2 -2mol%Y_2O_3 -1mol%Gd_2O_3 -1mol%Yb_2O_3– TransTech) and second, during this deposition the vapor flux was interrupted using the deposition shutter to create non-distinct interfaces. The deposition shutter was open for a period of 2.69 minutes and closed for 0.5 minutes and this was repeated for a total of 41 layers. The shutter was then closed and the electron beam guns were ramped down and turned off and the samples were allowed to cool for 10 minutes in the load lock before the chamber was vented.

Matrix 6

The EB-PVD coating chamber setup was modified to that shown in Figure 1 (b,c). A GZO ingot was placed in crucible D while a low-k YSZ ingot was in crucible B. The samples were positioned directly over crucible B, but offset left towards crucible D by ~0.5 inch. This was done to increase the incorporated vapor flux from ingot D (GZO). Sample heating to 1025 °C and TGO formation continued as described in Matrix 1. After a 20 minute soak, the melt was established on ingot B (low-k YSZ) via gun #6 and coating deposition was continued at a feed rate of 0.8 mm/min to obtain a low-k YSZ layer approximately 25 μm thick. Near the end of this 25 μm layer, the melt was established on ingot D (GZO) via gun #3. Ingot D was evaporated at a feed rate of 0.4 mm/min, while the low-k ingot was evaporated at a feed rate of 0.8 mm/min yielding a low-k to GZO ratio of 1:1. Note that the vapor shield in the chamber setup acts to separate the vapor clouds from each evaporated ingot to create a multilayer coating with distinct interfaces of alternating low-k and GZO layers. The final cool down portion of the run was similar as described for Matrix 1, with oxygen flow increased to 150 sccm and samples cooled to <500 °C before being translated back into the load-lock chamber. The shutter was then closed and the electron beam guns were ramped down and turned off and the samples were allowed to cool for 10 minutes in the load lock before the chamber was vented.

Matrix 7

The experimental details and procedures were identical to Matrix 6, with the exception that the low-k to GZO ratio was 1:4. This was obtained by changing the feed rate of both the low-k YSZ and GZO ingots. Thus, 36 mm of each ingot was fed at a feed rate of 0.6 mm/min. The shutter was then closed and the electron beam guns were ramped down and turned off and the samples were allowed to cool for 10 minutes in the load lock before the chamber was vented.

Fracture surfaces, cross-sections, and surface morphologies of the coated samples were examined by an FEI Quanta 200 environmental scanning electron microscope (ESEM) to determine microstructural and morphological differences within the various coatings. Phase analysis was determined using Philips X'Pert model MPD and MRD X-ray diffractometers. Thermal conductivity of the TBC was measured by the steady-state heat flux (CO_2 laser) technique at 1316 °C [17].

Particle erosion was performed using an in-house erosion rig. The TBC samples were loaded into the sample holder and mask. The sample holder was then mounted into the erosion rig directly beneath the particle acceleration tube. This ensured consistency through all the erosion trials and also allowed for adjustable incident angle of erodent. After mounting, a measured amount of media, in this case, 100 grams of 240 grit (~50 μm) Al_2O_3, was placed in the hopper. During testing, a pressurized feed system feeds the powder from the hopper into the acceleration tube, where compressed air accelerates the erosion particles. Air pressures of 30 PSI were used to accelerate particles to speeds of 100 m/s at incident angles of 30° and 90°. Samples were weighed before and after each erodent dose exposure to determine coating weight change which was plotted as a function of erodent fed. Coatings were considered to have failed when the substrate was clearly visible.

RESULTS AND DISCUSSION

X-Ray Diffraction

Figure 2 is an XRD pattern of each sample from Matrices 1-7. Due to the difficulty in distinguishing the phases, XRD scans were performed at 71-77 and 27-33 degrees 2θ and are shown Figures 3 and 4, respectively. The high angles were selected to differentiate the {400} reflections from the tetragonal, tetragonal prime and cubic phases [18,19], while the low angles confirmed the distinction. In Figure 3, the non GZO containing coatings from Matrices 1, 2 and 5 exhibit strong peaks at just over 74° 2θ. The pure GZO coatings from Matrices 3 and 4 are flat in this region and therefore the 74° peak is attributed to the t' phase. The opposite is true for the peaks just under 72° and so that peak is attributed to the cubic phase. Matrices 6 and 7 exhibit two weak peaks in Figure 3 that are likely both cubic and t' peaks which have been shifted due to strain from the layered structure. The low angle scan in Figure 4 further exacerbates the differences between the coatings. Matrices 1,2 and 5 (YSZ, low-k and multilayer low-k, respectively) all have a single distinct peak at just over 30° 2θ. These coatings contain no GZO, therefore the 30° peak is attributed to the t' phase of YSZ. Conversely, the coatings composed of pure GZO from Matrices 3 and 4 exhibit a peak at around 29.5 ° 2θ, while the pure YSZ coatings from Matrices 1,2 and 5 have no peak in this region. This peak is therefore attributed to the cubic phase of GZO and similar to the high angle scan, both peaks are exhibited by Matrices 6 and 7, confirming the composite multilayered nature of the coatings.

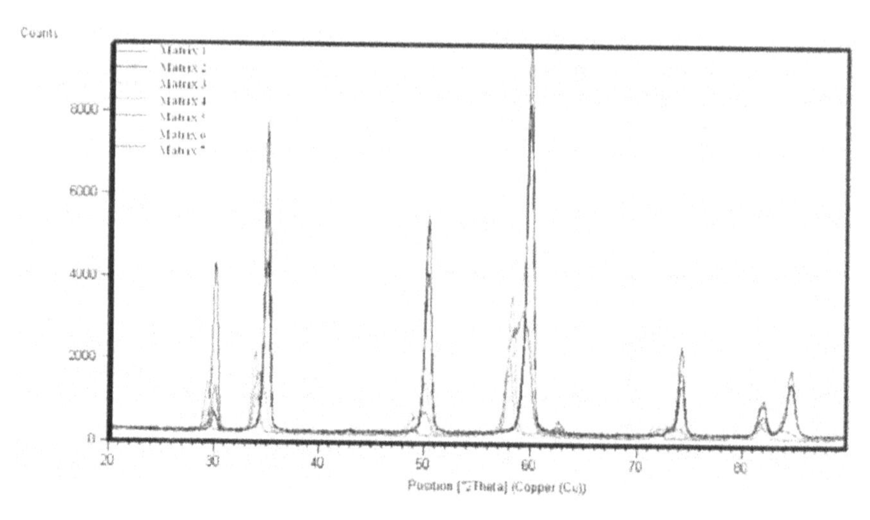

Figure 2: XRD scan from 20-90 degrees 2θ for samples from Matrices 1-7. Notice the peaks are difficult to distinguish and thus phase confirmation is unclear.

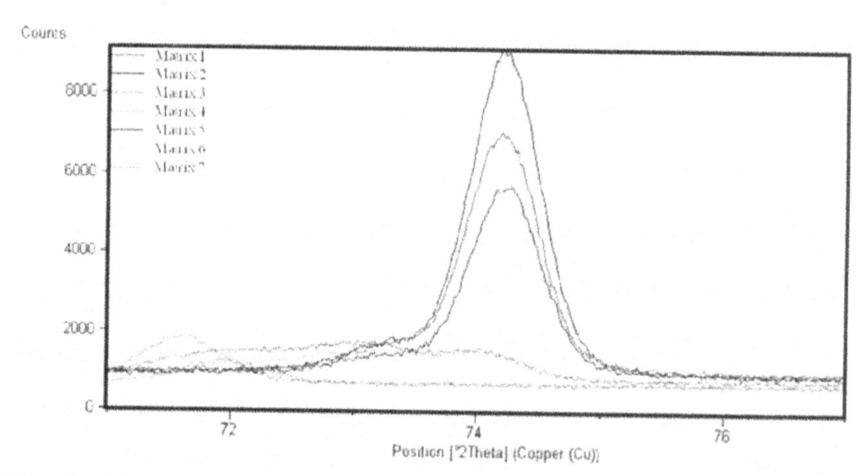

Figure 3: XRD scan from 71-77 degrees 2θ for samples from Matrices 1-7. Matrices 1, 2 and 5 have large peaks at just over 74 degrees 2θ, indicating a t' diffraction peak. Matrices 3 and 4 are flat in this region, confirming they have no t' phase and are purely cubic. Matrices 6 and 7 show a mixture of phases, corresponding to the low-k—GZO multilayer composite.

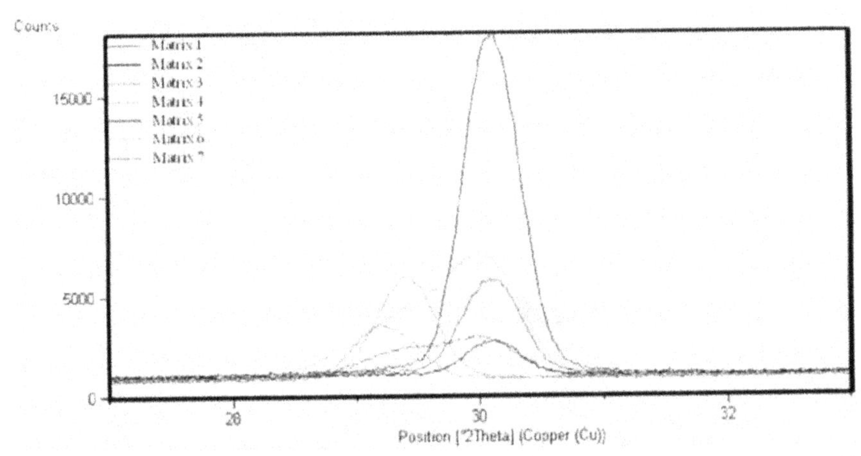

Figure 4: XRD scan from 27-33 degrees 2θ for samples from Matrices 1-7. Matrices 1, 2 and 5 have large peaks at just over 30 degrees while Matrices 3 and 4 are flat in this region. This confirms that Matrices 3 and 4 are composed of a cubic phase while Matrices 1, 2 and 5 are composed of a t' phase. Also, Matrices 6 and 7 show a mixture of both cubic and t', confirming low-k—GZO composite composition

Microscopy

Figures 5-9 are electron micrographs of the sample surfaces and cross sections for the coatings from Matrices 1-7. The coatings all exhibit the columnar microstructure characteristic of EB-PVD TBCs. Figure 5 (a) shows the surface morphology of the Matrix 1 YSZ coating, while Figure 5 (b) is a cross sectional micrograph showing the intercolumnar porosity and columnar microstructure. Figures 6 (a) and (b) show the surface morphology and cross section, respectively, of the low-k YSZ sample from Matrix 2. The morphology and intercolumnar porosity appear similar to the YSZ from Matrix 1 in Figure 5. Figure 6 (c) shows the surface morphology of the 41 layer low-k sample from Matrix 5. The surface morphology shown in Figure 6 (c) appears to have fewer of the small features seen in the single layer coating surface in Figure 6 (a). The vertical intercolumnar porosity appears to remain unchanged in the 41 layer low-k {Figure 6 (d)} when compared to the single layer low-k {Figure 6 (b)}. The microstructure does appear to have changed, with more continuous columns in the 41 layer low-k in Figure 6 (d) and more discontinuity in columns of the single layer low-k in Figure 6 (b).

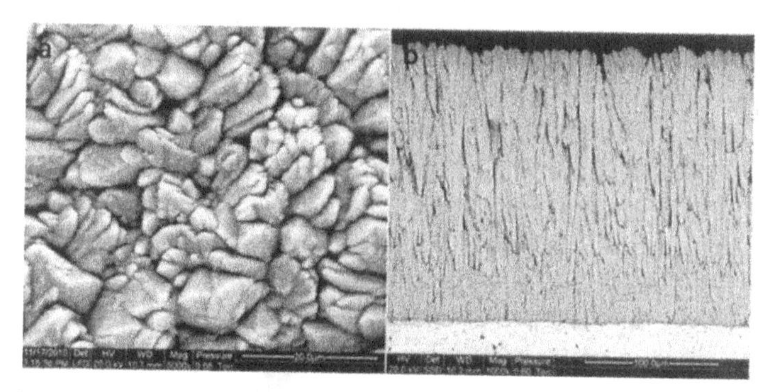

Figure 5: SEM micrograph of the YSZ coating from Matrix 1 showing the: (a) surface morphology and (b) backscattered cross sectional image.

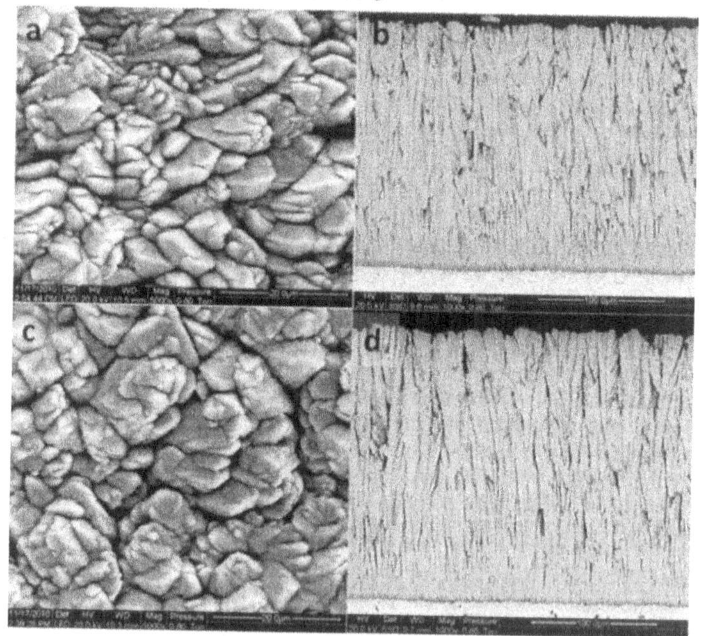

Figure 6: SEM micrographs of low k YSZ showing the: (a) surface morphology and (b) backscattered cross sectional image of the single layer low-k coating from Matrix 2, (c) surface morphology and (d) backscattered cross sectional image of the multilayer low-k coating from Matrix 5.

SEM micrographs of the GZO coatings from Matrices 3 and 4 are presented in Figure 7. The surface morphologies shown in Figure 7 (a) and (c) are drastically different than those of the YSZ and low-k shown in Figure 5 (a) and Figure 6 (a,c). This is likely a texturing phenomena exhibited by the cubic GZO which grew in the [111] direction. The [111] direction is the fast growing direction perpendicular to the close packed (111) plane, hence the angular column tips in Figure 7 (a) and (c). The GZO cross sections shown in Figure 7 (b) and (d) also exhibit a significant degree of intercolumnar porosity when compared to those of YSZ and low-k in Figure 5 (b) and Figure 6 (b,d).

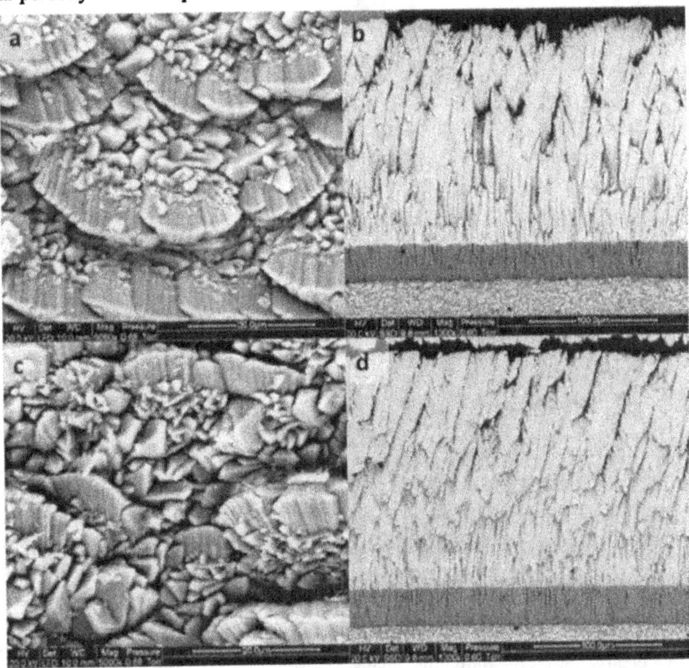

Figure 7: SEM micrographs of GZO (a/b) and 40 layer GZO (c/d) showing the: (a) surface morphology and (b) backscattered cross sectional image of the GZO coating from Matrix 3, (c) surface morphology and (d) backscattered cross sectional image of the low-k coating from Matrix 4. Notice the features of the 40 layer GZO coating in (c) appear larger than those of the single layer GZO in (a).

Also, the columns have grown at an angle, possibly due to an operation error in which the samples were not translated directly above the GZO ingot, differences in nucleation, growth and strain energy, or variation in the vapor cloud. The surface features of the single layer GZO coating in Figure 7 (a) appear finer than those of the 40 layer GZO coating in Figure 7 (c). This same trend was observed for the low-k/41 layer low-k samples. It appears that the shutter method has effectively allowed the samples to diffuse (during shuttering) and thus nucleate fewer columns and therefore larger features are seen on the surface. The low-k—GZO multilayer composites from Matrices 6 and 7 are shown in Figures 8 and 9, respectively. The surface morphologies shown in Figure 8 (a) and 9 (a) are similar to those of the GZO coating from Figure 7. One difference is that the roughness appears lower for the multilayer composites from Matrices 6 and 7 than that for the GZO samples from Matrices 3 and 4. Also, the column tips are fairly smooth in the GZO coatings in Figure 7, while the multilayer composite columns tips in Figure 8 and 9 have small "growths" on them. The cross sections in Figure 8 and 9 (b-d) are similar to those of the YSZ and low-k from Figures 5 (b) and 6 (b,d), with intercolumnar porosity. In Figure 9 (b) it appears that the coating has lighter and darker regions.

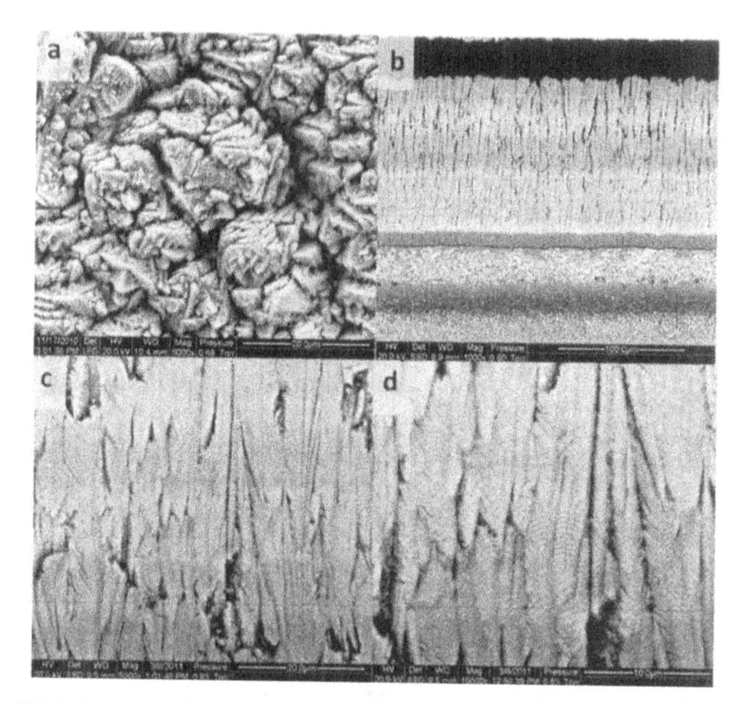

Figure 8: SEM micrographs of the low-k—GZO multilayer composite from Matrix 6 showing: (a) surface morphology, (b) a backscattered cross sectional image of the TBC system, (c) higher magnification view of the columnar structure where the low-k—GZO layers become visible, and (d) micrograph that clearly shows the individual layers.

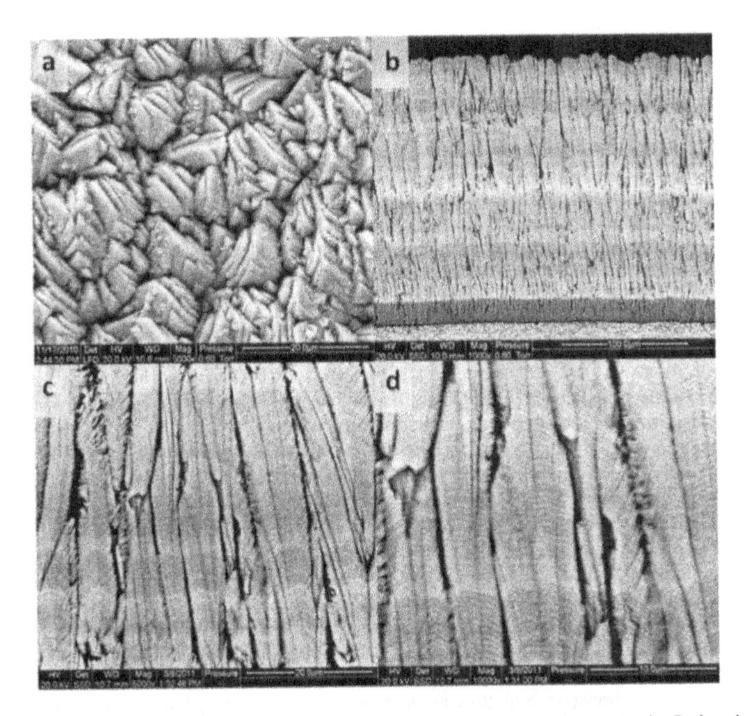

Figure 9: SEM micrographs of the low-k—GZO multilayer composite from Matrix 7 showing: (a) surface morphology, (b) a backscattered cross sectional image of the TBC system, (c) higher magnification view of the columnar structure where the low-k—GZO layers become visible, and (d) micrograph that clearly shows the individual layers. Note the dark band in (c) and (d) which indicates the local thickness increase of the low-k layers have increased.

Figures 9 (c) and (d) verify the presence of a dark band. It appears the local thickness of the darker layer (low-k) has increased and thus there is a dark band. The bright band in the middle of Figure 9 (b) is likely a local increase in the thickness of the GZO layers with respect to low-k layer thickness. The low-k—GZO multilayer composite coatings have surface morphologies similar to the cubic GZO, but columnar structure and a degree of porosity more similar to YSZ and low-k.

Thermal Conductivity

Thermal conductivity results of all the fabricated TBC systems (Table 1) are shown in Figure 10. The monolayered YSZ sample showed the higher initial or "as deposited" thermal conductivity of ~1.45 W/m-K, along with the highest 20 hour sintered thermal conductivity of 2.2 W/m-K. The monolayered low-k and GZO coatings have decreased initial thermal conductivities of 1.2 and 1.13 W/m-K, respectively, along with a decreased rate of increase of thermal conductivity with time, with 20 hour sintered thermal conductivities of 1.64 and 1.42 W/m-K, (Figure 10). In the case of GZO, layering with non-distinct layers has shown to effectively decrease the thermal conductivity, with an initial thermal conductivity of 1.08 W/m-K and a 20 hour sintered thermal conductivity of 1.33 W/m-K. Interestingly, though the layering effect served to decrease the initial thermal conductivity, it can be argued that layering had little effect on the sintering behavior of GZO. Both GZO coatings had a ~23-26% increase of thermal conductivity (from initial to max) at 1316 °C surface temperatures. This increase of thermal conductivity over time is still much less than the ~52% experienced by YSZ. Layering with non-distinct layers did not decrease the thermal conductivity of the low-k coatings. The layered low-k coating had an initial thermal conductivity of 1.33 W/m-K and a 20 hour sintered thermal conductivity of 1.62 W/m-K. Unlike GZO, layering has shown a reduction in sintering rate for low-k, with a ~37% increase in thermal conductivity for the single layer coating and a 22% increase in the layered coating. The microstructures examined previously may explain some of the thermal conductivity trends. The GZO may have had the lowest initial thermal conductivity due to the angular column growth shown in Figure 7 (b,d) which led to a more significant contribution of the intercolumnar porosity in reducing thermal conductivity. Also, the 41 layer low-k sample (Matrix 5) had decreased sintering compared to the monolayered low k (Matrix 2). Figure 6 (d) shows that the Matrix 5 sample had more continuous columns than the Matrix 2 sample {Figure 6 (b)}. Essentially there is a greater propensity of for sintering in the monolayer low k coating.

The multilayer GZO/low-k coating with distinct layers from Matrices 6 and 7 results are shown at the top of Figure 10. The thermal conductivities of Matrix 6 and 7 were found to be higher than their constituent single layers viz. low k and GZO. The thermal conductivities of the two multilayered GZO/low-k coatings were expected to be in the range of 1.13 (GZO) and 1.2 (low k). The observed higher thermal conductivities of the two multilayered coatings with distinct layers is most likely due to increased coating densities of the non-optimized microstructure, though this remains to be tested. Matrix 6 exhibited a thermal conductivity slightly less than layered low k with non-distinct layers, while Matrix 7 had a relatively large thermal conductivity. The increase in thermal conductivity from Matrix 6 to 7 is contrary to what was expected since Matrix 7 has a larger volume fraction of the lower thermal conductivity GZO. This can be explained by the increased columnar width and less interrupted microstructure of Matrix 7 (see Figure 9 vs. figure 8). This microstructure allows for an increased in phonon transport and thus an increase in conductivity, with respect to the Matrix 6 coating.

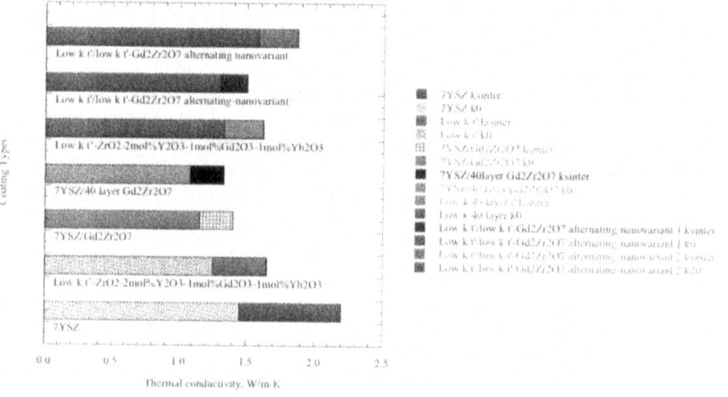

Figure 10: Plot of initial thermal conductivity and maximum thermal conductivity for the TBC systems from Matrices 1-5. The lowest conductivity was achieved with the 40 layer GZO sample, which also showed the least amount of sintering.

Particle Erosion

Particle erosion was performed at 30° and 90° incident angles on each sample from 1 (Matrices 1-7). Coatings from Matrices 1, 2 and 5 have shown superior erosion performance over the rest of the coatings. The erosion results are plotted in Figures 11 and 12 in terms of coating mass loss of coating versus mass of erodent used. A steady state erosion rate can be determined by the slope of the plots and is shown for each coating under both impingement angles in Figure 13.

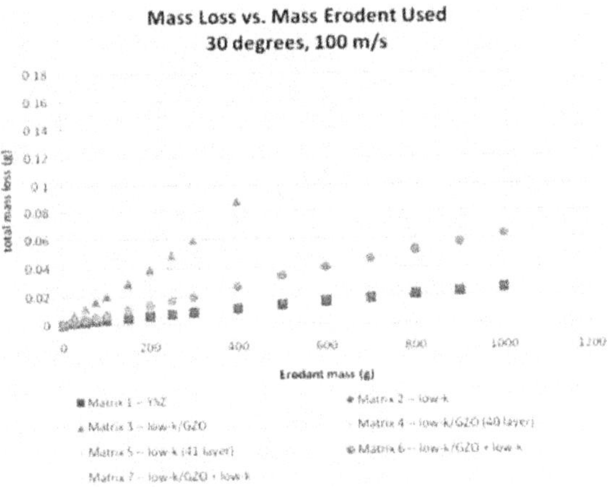

Figure 11: Erosion performance of coatings from Matrices 1-7 using 240 grit alumina media at 100 m/s velocities and 30° incidence. The GZO samples from Matrices 3 and 4 have the highest erosion rate while the YSZ and low-k YSZ samples from Matrices 1, 2 and 5 have the lowest erosion rate.

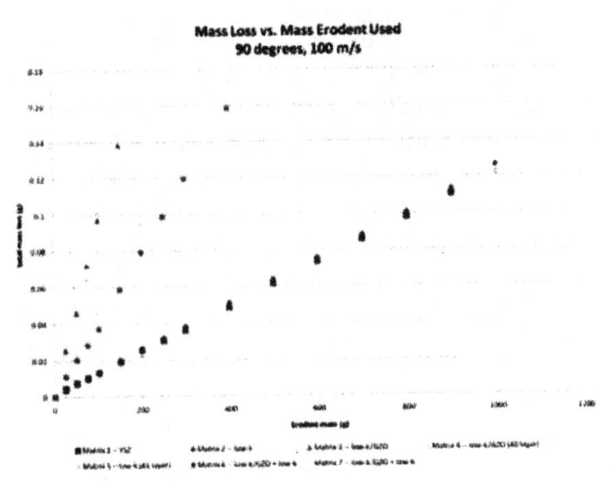

Figure 12: Erosion performance of coatings from Matrices 1-7 using 240 grit alumina media at 100 m/s velocities and 90° incidence. Though erosion rates were markedly increased, the samples performed in the same order as the 30° test with GZO having a high erosion rate while YSZ and low-k had the lowest erosion rates.

Figures 11 and 12 show that indeed the samples from Matrices 1, 2 and 5 (YSZ, low-k, 41 layer low-k) perform the best, with erosion rates of ~25 μg/g at 30° and ~125 μg/g at 90°, respectively. By comparing the 30° erosion performance data to the 90° erosion performance data, the effect of incidence angle on erosion for a ceramic TBC is clearly observed. This same trend has been shown in the literature [20]. Figure 13 shows that, for a given coating, the erosion rate increases by a factor of ~5-6X when the erodent impingement angle is increased from a 30° incident to a 90° incident. Also, the erosion rates of the GZO coated samples were ~7-8 times greater than that of the YSZ and low-k YSZ samples. The higher erosion rate for the GZO samples is attributed to several factors. First, the cubic structure of GZO has lower fracture toughness than the t' structure of YSZ and low-k YSZ. Therefore, for a given particle impact, it is more likely that GZO columns will crack and consequently, that more columns will crack. If more columns crack, then there is an increased probability that the crack is in a region where the TBC columns are not separated by porosity. If columns are touching, the crack can propagate through multiple columns. So, for a given particle impact on GZO, there is a larger effected region than in YSZ. Over time, these effected coating regions are spalled off, resulting in GZO having a higher erosion rate. Since the coatings from Matrices 6 and 7 contain both GZO and low-k, they have intermediate erosion rate. As expected, the higher GZO containing coating from Matrix 7 has an erosion rate greater than the coating from Matrix 6.

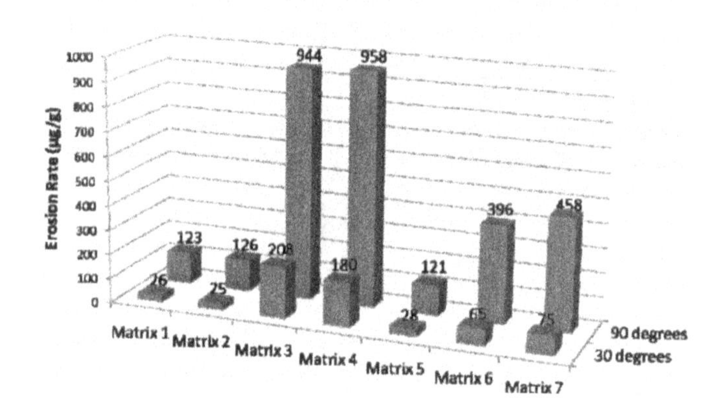

Figure 13: Plot of erosion rate for each of the Matrices at 30° and 90°. For a given coating, the 90° incident erosion rate is ~5X the 30° incident erosion rate. Also, the GZO coated samples from Matrices 3 and 4 range from ~3-8X the erosion rate of the other samples.

CONCLUSION

A variety of unique TBC architectures were deposited utilizing a combination of 7YSZ, low-k and GZO feed materials. The coatings exhibited the columnar microstructure characteristic of EB-PVD TBCs. The low-k coatings exhibited room temperature erosion performance similar to that of YSZ, while coatings containing GZO had an 8X increase in erosion rate. The poor erosion rate of the pure GZO coatings was likely due to the cubic structure as well as the angular growth of the columns. The multilayer low-k/GZO coatings exhibited an intermediate erosion rate between that of GZO and low-k, with the larger volume fraction low-k coating outperforming the lower volume fraction low-k coating. The increase in performance with respect to low-k volume fraction is attributed to the volume increase of the tougher t' phase of the low-k YSZ. The thermal conductivity of both the pure GZO and low-k coatings was markedly lower than that of 7YSZ. While YSZ measured an initial thermal conductivity of 1.45 W/m-K, pure GZO measured an value of 1.13 W/m-K which was lowered even further to 1.08 W/m-K through the use of layering. The increase in conductivity over time, and therefore sintering, was reduced by as much as 15% for the multilayer low-k coating. Therefore, it is apparent that a unique multilayer architecture can allow for reduced thermal conductivities, an increased sintering resistance and increased erosion performance.

ACKNOWLEDGMENTS

This research was sponsored by the Department of Energy (DOE) STTR under award number DE-SC0004356. Any opinions, findings, conclusions, or recommendations expressed in this material are those of the authors and do not necessarily reflect the views of the US DOE.

REFERENCES

[1] Zhang, Y., Haynes, J. A., Pint, B. A., Wright, I. G., & Lee, W. Y. (2003). Martensitic transformation in CVD NiAl and (Ni,Pt)Al bond coatings. *Surface and Coatings Technology*, *163-164*, 19-24.

[2] Chen, M. W., Ott, R. T., Hufnagel, T. C., Wright, P. K., & Hemker, K. J. (2003). Microstructural evolution of platinum modified nickel aluminide bond coat during thermal cycling. *Surface and Coatings Technology*, *163-164*, 35-30.

[3] Darzens, S., Mumm, D. R., Clarke, D. R., & Evans, A. G. (2003). Observations and analysis of the influence of phase transformations on the instability of the thermally grown oxide in a thermal barrier system. *Metallurgical and Materials Transactions A*, *34*, 511-522.

[4] Deodeshmukh, V., Mu, N., Li, B., & Gleeson, B. (2006). Hot corrosion and oxidation behavior of a novel Pt + Hf modified y'-Ni3Al + y-Ni based coatings. *Surface and Coatings Technology*, *201*, 3836-3840.

[5] Das, D. K., Murphy, K. S., Ma, S., & Pollock, T. M. (2008). Formation of Secondary Reaction Zones in Diffusion Aluminide-Coated Ni-Base Single-Crystal Superalloys Containing Ruthenium. *Metallurgical and Materials Transactions A*, *39*, 1647-1657.

[6] Levi, C. G. (2004). Emerging materials and processes for thermal barrier systems. *Current Opinion in Solid State and Materials Science*, 77-91.

[7] Vassen, R., Cao, X., Tietz, F., Basu, D., & Detlev, S. (2000). Zirconates as New Materials For Thermal Barrier Coatings. *Journal of the American Ceramic Society*, *83* (8), 2023-2028.

[8] Maloney, M. J. (2000). *Patent No. 6,117,560.* United States.

[9] Wu, J., Wei, X., Padture, N., Klemens, P., Gell, M., Garcia, E., et al. (2002). Low-Thermal Conductivity Rar Earth Zirconates for Potential Thermal Barrier Coating Applications. *Journal of the American Ceramic Society*, *85* (12), 2031-2035.

[10] Kramer, S., Yang, J., & Levi, C. G. (2008). Infiltration inhibiting reaction of gadolinium zirconate thermal barrier coatings with CMAS melts. *Journal of the American Ceramic Society*, *91* (2), 576-583.

[11] Vaben, R., Jarligo, M. O., Steinke, T., Mack, D. A., & Stover, D. (2010). Overview on advanced thermal barrier coatings. *Surface and Coatings Technology*, *205*, 938-9412.

[12] Zhao, H., Begley, M. R., Heuer, A., Sharghi-Moshtaghin, R., & Wadley, H. N. (2011). Reaction, transformation and delamination of samarium zirconate thermal barrier coatings. *Surface and Coatings Technology*, *205*, 4355-4365.

[13] Zhu, D., & Miller, R. A. (2002). Thermal Conductivity and Sintering Behavior of Advanced Thermal Barrier Coatings. In H.-T. Lin, & M. Singh (Ed.), *26th Annual Conference on Composites, Advanced Ceramics, Materials, and Structures: B: Ceramic Engineering and Science Proceedings. 23*, pp. 457-468. Hoboken NJ: John Wiley & Sons, Inc.

[14] Zhu, D., Chen, Y. L., & Miller, R. A. (2004). *Defect slutering and nanophase structure characterization of multicomponent rare earth oxide doped zirconia-yttria thermal barrier coatings.* Glenn Research Center, Cleveland: NASA Technical Memo 212480.

[15] Zhu, D., Nesbitt, J. A., McCue, T. R., Barrett, C. A., & Miller, R. A. (2002). Furnace Cyclic Behavior of Plasma-Sprayed Ziconia-Yttria and Multi-Component Rare Earth Oxide Doped Thermal Barrier Coatings. *Ceramic Engineering and Science Proceedings. 23*, pp. 533-545. The American Ceramic Society.

[16] Wolfe, D. E., Singh, J., Miller, R. A., Eldridge, J. I., & Zhu, D.-M. (2005). Tailored microstructure of EB-PVD 8YSZ thermal barrier coatings with low thermal conductivity and high thermal reflectivity for turbine applications. *Surface and Coatings Technology* , 132-149.

[17] Zhu, D., Miller, R. A., Nagaraj, B. A., & Bruce, R. W. (2001). Thermal conductivity of EB-PVD thermal barrier coatigns evaluated by a steady state laser heat flux technique. *Surface and Coatings Technology* , 1-8.

[18] Lughi, V., & Clarke, D. R. (2005). High temperature aging of YSZ coatings and subsequent transformation at low temperature. *Surface and Coatings Technology* , 1287-1291.

[19] Lughi, V., & Clarke, D. R. (2005). Transformation of electron beam physical vapor-deposited 8 wt% yttria-stabilized zirconia thermal barrier coatings. *Journal of the American Ceramic Society* , 2552-2558.

[20] Wellmann, R. G., & Nicholls, J. R. (2007). A review of the erosion of thermal barrier coatings. *J. Phys. D: Appl. Phys* , R293-R305.

FOREIGN OBJECT DAMAGE (FOD) BEHAVIOR OF EB-PVD THERMAL BARRIER COATINGS (TBCs) IN AIRFOIL COMPONENTS

Jennifer Wright, D. Calvin Faucett, Matt Ayre, Sung R. Choi*
Naval Air Systems Command, Patuxent River, MD 20670

ABSTRACT

A series of foreign-object-damage (FOD) tests on ceramic thermal barrier coatings (TBCs) of aeroengine airfoil components were performed to determine their responses to ballistic particle impact. Airfoils coated with electron beam, physical vapor deposition (EB-PVD) TBCs were FOD-tested by 1.6 mm-diameter spherical projectiles of two different materials of borosilicate glass and silicon nitride. A range of impact velocities from 150 to 300 m/s was employed for each of projectile materials. Degree and morphologies of impact damage were quantified and characterized in terms of impact velocity, projectile material, and component service-life. A first-order approximation of impact force was made based on the energy balance principle and the 'contact yield pressure' analysis.

INTRODUCTION

The brittle nature of ceramic materials, either monolithic ceramics or ceramic matrix composites (CMCs), has raised concerns about structural damage when they are used as aeroengine components and are subjected to impact by foreign objects. This has prompted the propulsion communities to take into account foreign object damage (FOD) as an important design parameter. A large amount of work on impact damage of monolithic ceramic materials has been done in the past decades [1-15]. Also, work on ballistic impact in CMCs, although sparse, has continued to include different types of CMC material systems [16-25]. Ceramic environmental barrier coatings (EBCs) in SiC/SiC composites were also assessed in their responses to ballistic impact [26,27]. The work on ceramic armor, although confined in armor applications, has also advanced our understanding of ballistic impact in terms of mechanisms, materials, and characterizations both experimentally and analytically [e.g., 28-31].

Thermal barrier coatings (TBCs) have been a prerequisite for hot-section components of advanced gas turbines because of their ability to provide thermal insulation to related metallic engine components. The merits of using ceramic TBCs are well recognized as a means of substantial increase in engine operating temperature with reduced cooling requirements, resulting in significant improvements in thermal efficiency, performance, and reliability. The zirconia-based ceramics, typically 7-8 wt% Y_2O_3-ZrO_2, are the most important and widely-utilized coating materials because of their low thermal conductivity, substantial strain tolerance, and unique microstructure achieved via plasma spraying or electron beam, physical vapor deposition (EB-PVD) process. However, the durability of TBCs under severe thermal and mechanical loading encountered in heat engines remains one of the major issues, primarily attributed to lower mechanical properties of the coatings [32]. This leads to another challenging issue on their significant susceptibility to impact damage by foreign objects ingested into engines or by particles torn from other engine components or by combustion products. A considerable amount of work has been done on the subjects of erosion on TBCs [e.g., 33-36]. However, despite its significance, work on FOD in TBCs has been relatively sparse particularly in aeroengine components [37-41].

The current paper presents the results of experimental work on FOD behavior of TBCs in actual aeroengine airfoil components. The airfoil components coated with 7-8wt% Y_2O_3-ZrO_2 (7-

* Corresponding author; Email address: sung.choi1@navy.mil

8YSZ) topcoats by EB-PVD were subjected to ballistic impact by 1.6 mm spherical ball projectiles at impact velocities of 100 to 300 m/s. Two different projectile materials, borosilicate and silicon nitride, were employed. The previously obtained data with steel ball projectiles [42] were used for comparison. Foreign object damage was characterized and analyzed through damage morphologies and Weibull statistics with respect to impact velocity, projectile material, and component service life. A first-order approximation of impact force was made and its validity was assessed via experimental damage data.

EXPERIMENTAL PROCEDURES

Materials/Components

The airfoils used in this work were high-pressure turbine (HPT) components coated with EB-PVD TBCs as topcoats. The TBCs were 7wt% YSZ. The coating thickness was about 110 μm. The bondcoats were Pt-Al with a thickness of about 60 μm. The substrates were Ni-based superalloy. The HPT airfoils were from three different aeroengines having their respective 'service lives' (hours) of 3, 14, and 22.[†] The detailed descriptions of the topcoats, the bondcoats, and the substrate are proprietary in nature and thus limited in this paper.

Foreign-Object-Damage (FOD) Testing

A ballistic impact gun, as described elsewhere [13,16], was utilized to conduct FOD testing. Two different materials of projectiles, borosilicate glass and silicon nitride, were used and their basic physical and mechanical properties are shown in Table 1. A spherical projectile with a diameter of 1.6 mm was inserted into a 300mm-long gun barrel. Helium gas and relief valves were utilized to pressurize and regulate a reservoir to a specific level, depending on prescribed impact velocity. Upon reaching a specific level of pressure, a solenoid valve was instantaneously released accelerating the projectile through the gun barrel to impact onto TBCs in the suction side of a target airfoil. A total of three different impact velocities of 150, 200 and 300 m/s were used for a given projectile material. For a given impact velocity, four impacts were applied in a random but systematic manner, away from the both leading and trailing edges, as shown in Fig. 1. All impact testing was conducted in a normal incidence angle. Two airfoils were chosen per projectile material and per aeroengine service hour, using a total of 12 airfoils all together for the whole experiments.

Table 1. Basic properties of projectile materials used in this work

Projectiles	Density[#] ρ (g/cm^3)	Vickers hardness[@] H_v (GPa)	Elastic modulus[^] E (GPa)	Fracture toughness[^] K_{IC} (MPam$^{1/2}$)
Borosilicate glass	2.40(0.02)	5.1(0.9)	62	0.7
Silicon nitride	3.21(0.02)	15.1(0.2)	320	5.5
Chrome steel (SAE52100)	7.78(0.05)	8.2(0.2)	200	15-20

Notes:
[#] By mass/volume method with five projectiles used
[@] By Vickers micro-indentation with an indentation load of 9.8N (ASTM C 1327 & E 384)
[^] From literature data
The numbers in the parentheses represent ±1.0 standard deviation.

[†] The service life given is in convention a characteristically relative numerical quantity associated with a specific life-controlling operation but not an absolute (total) service hour. A life of '23' is referred to as a life limit of the component concerned in this work.

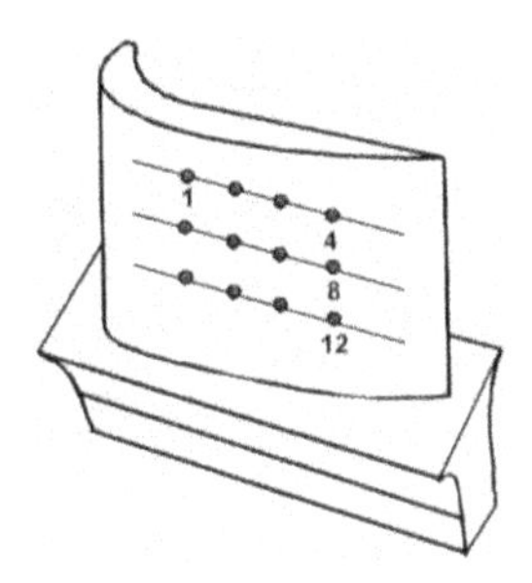

Figure 1. A schematic showing the locations of impact in a HPT EB-PVD airfoil applied in the FOD testing.

Impact Morphologies

Impact morphologies were examined from the impact sites and in some cases from the cross-sections, using optical and/or scanning electron microscopy.

RESULTS AND DISCUSSION

Impact by Borosilicate Glass Projectiles

Figure 2 shows the impact sites generated in the EB-PVD HPT airfoils at V=150, 200, and 300 m/s by borosilicate glass projectiles. Regardless of impact velocity, all the impact sites showed only circular impressions with some traces of ring cracks along the impression peripherals. A summary of the impact-damage size measurements is presented in Fig. 3, where impact damage size (diameter, 'd') was plotted against impact velocity for three different service lives of the related airfoil components. The damage size increased monotonically with increasing impact velocity and did not exhibit any significant effect on the component service life. The nature of monotonic increase in damage size with increasing impact velocity is also depicted in a Weibull distribution, as shown in Fig. 4, in which the probability of occurrence (F) was plotted in a two-parameter Weibull scheme with respect to damage size for all the component service lives. The occurrence, showing a uni-modal distribution with a Weibull modulus (the slope) of ~10, implies that the impact damage generated by the borosilicate glass projectiles would have been under the almost identical damage mechanism in a whole range of applied impact velocities. This would be more self-evident when compared with the silicon nitride data, which will be described in the next section.

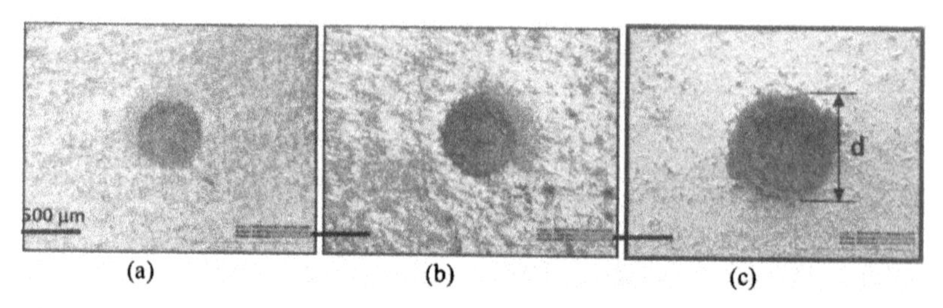

Figure 2. Typical appearances of impact damage for EB-PVD TBCs in HPT airfoils impacted by 1.6-mm diameter borosilicate glass ball projectiles at: (a) V=150 m/s, (b) V=200 m/s, and (c) V=300 m/s. Bar: 500μm.

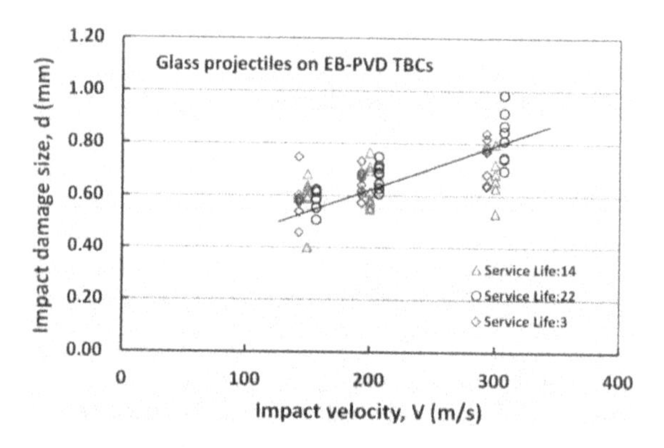

Figure 3. External impact damage size as a function of impact velocity with different service lives (hours) of EB-PVD TBC airfoils impacted by 1.6-mm diameter borosilicate glass ball projectiles.

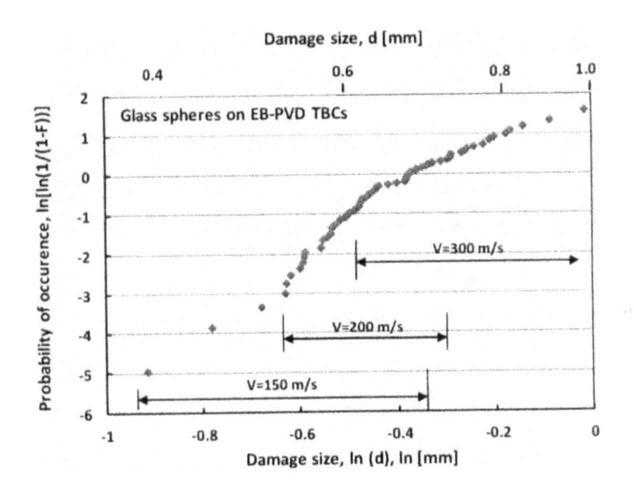

Figure 4. Two-parameter Weibull distribution of damage sizes for EB-PVD TBC airfoils impacted at V=150, 200, and 300 m/s by 1.6-mm diameter borosilicate glass ball projectiles. The range of damage size corresponding to respective impact velocities is indicated in the plot.

Impact by Silicon Nitride Projectiles

Figure 5 shows a typical impact damage generated by silicon nitride ball projectiles. At lower impact velocities of V=150 and 200 m/s, the impact damage showed primarily circular impressions with some vague signs of delamination of the coatings. At V=300 m/s, the damage was significant in a form of spallation of TBCs with a presence of subsurface coating delamination. The spallation occurred from the TBCs/bondcoat interfaces (presumably from TGO layer). The results of the impact-damage size measurements are shown in Fig. 6. The damage size increased slightly from 150 m/s to 200 m/s, similar to the cases for the borosilicate glass projectiles. However, at V=300 m/s, substantial increase in damage size resulted in, accompanying the spallation of TBCs. A significant data scatter, as seen from in Fig. 6, is also noted at V=300 m/s.

A two-parameter Weibull distribution of impact damage size for all the data in Fig. 6 is illustrated in Fig. 7. A distinctive bimodality is typified of the distribution, such that Region 'A' is related to the lower impact velocities of V=150 and 200 m/s generating impression only, while Region 'B' is related to the higher impact velocity of V =300 m/s resulting in spallation. Weibull modulus was ~12 and ~1, respectively, for Regions 'A' and 'B.' The Weibull modulus in Region 'A' is close to that (~10) of the borosilicate glass projectiles, indicating a similar mode of damage having occurred by both glass and silicon nitride projectiles. The considerable data scatter at V=300 m/s indicates again the stochastic nature of impact responses to the TBCs. This requires particular attention that sample size should be enough to ensure statistical reproducibility and reliability, as having observed from many brittle materials (monolithic ceramics or CMCs) [13-16]. Use of only a few data points should be avoided and is strongly discouraged.

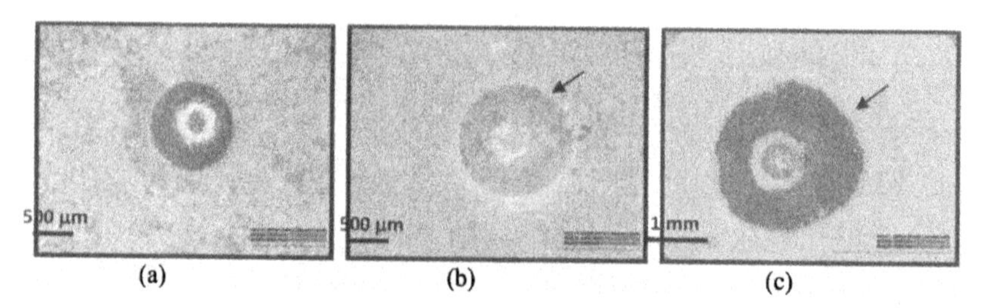

Figure 5. Typical appearances of impact damage for EB-PVD TBCs in HPT airfoils impacted by 1.6-mm diameter silicon nitride ball projectiles at: (a) V=150 m/s, (b) V=200 m/s, and (c) V=300 m/s. Bar: 500μm. The arrows represent regions (or shadows) of subsurface delamination.

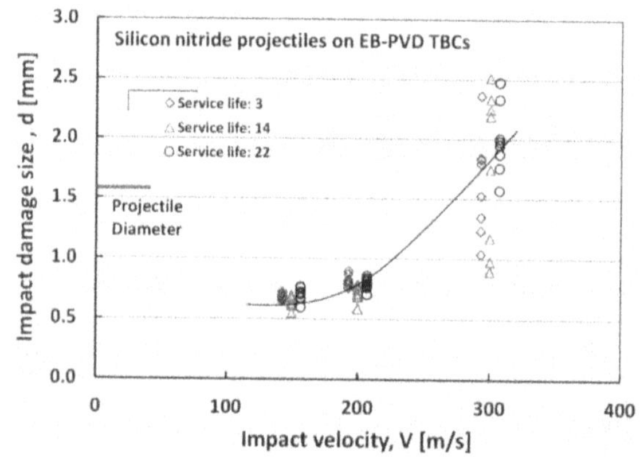

Figure 6. External impact damage size as a function of impact velocity with different service lives (hours) of EB-PVD TBC airfoils impacted by 1.6-mm diameter silicon nitride ball projectiles.

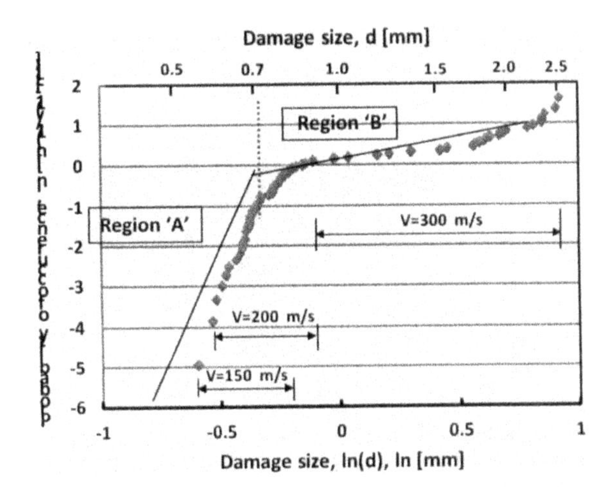

Figure 7. Two-parameter Weibull distribution of damage sizes for EB-PVD TBC airfoils impacted at V=150, 200, and 300 m/s by 1.6-mm diameter borosilicate glass ball projectiles. The range of damage size corresponding to respective impact velocities is indicated in the plot.

Cross-Sectional Impact Morphology

Figure 8 shows a cross-sectional view of impact damage at V=150 m/s by borosilicate glass projectiles. Plastic deformation of topcoat, bondcoat, and substrate is clearly seen, with some sign of minor densification of the TBCs. Also, delamination occurred beneath the top coat, along the topcoat/bondcoat interface. Figure 9 shows a cross-section of impact damage at V=200 m/s in which several features of the damage are noted including plastic deformation, densification of the TBCs, shear band (or kinking or cone cracking) [35], and delamination. In particular, the delamination crack was almost three times greater in size than the impact impression. Such predominant delamination cracking was also observed at the highest impact velocity of V=300 m/s, where significant TBCs spallation was also accompanied, as seen in Figure 10. The delamination was not symmetric with respect to the impact site, due to some variations in coating thickness; however, its delamination crack size was about three times greater than the spallation size.

A summary of the delamination crack-size measurements is presented in Fig.11, where the delamination data was compared with the external damage counterpart (Figs. 3 and 6). The borosilicate glass projectiles resulted in delamination about twice larger than the external impact damage size; whereas, the silicon nitride projectiles generated two to three times greater than the external damage size. A similar behavior was also observed from the previous work on plasma-sprayed TBCs impacted by 1.59 mm-diameter steel projectiles [42]. The results shown in this section indicate an importance of appropriate assessment tools/methods that must be exercised for accurate damage-morphology characterization, thus enabling to provide better ways to tailoring coating systems, materials redesign, modeling, life prediction, and so on. It should be mentioned here that the delamination data presented in this work was limited with only a few data points; so more cross-section work needs to be done for improved reproducibility of the overall data. Furthermore, prediction of delamination cracking with respect to impact variables (velocity, projectile materials) should be sought using appropriate delamination models [35, 43,44], which would be a future task to conduct.

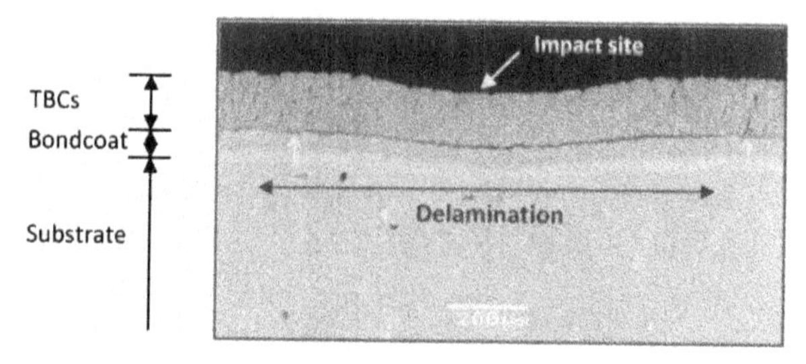

Figure 8. A typical example of the cross-sectional view of an EB-PVD TBC airfoil impacted at 150 m/s by 1.6-mm diameter borosilicate glass ball projectiles. Both impact site and delamination crack are seen in the figure. Service life: '14'.

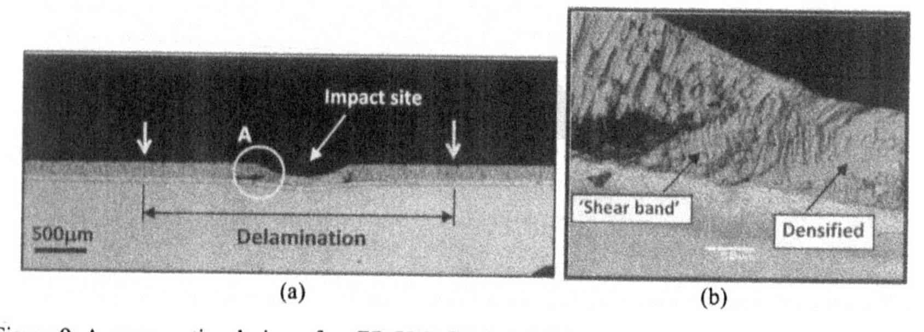

Figure 9. A cross-sectional view of an EB-PVD TBC airfoil impacted at 200 m/s by 1.6-mm diameter silicon nitride ball projectiles: (a) Overall view showing both impact site and delamination; (b) Details in Region 'A' showing 'shear band' (deformation) and densified region by an impacting projectile. Service life: '3.'

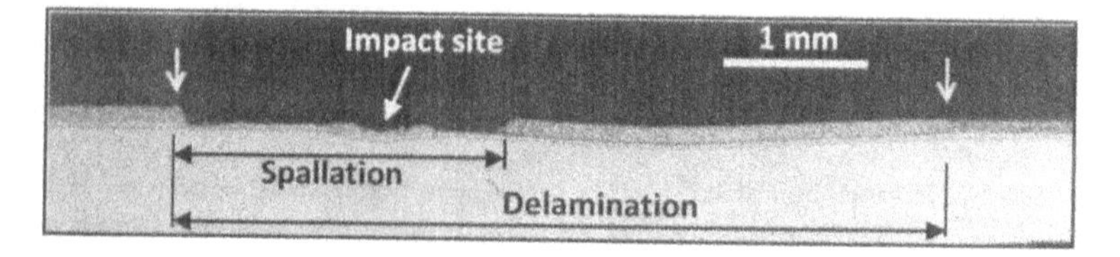

Figure 10. A cross-sectional view of an EB-PVD TBC airfoil impacted at 300 m/s by 1.6-mm diameter silicon nitride ball projectiles. Features such as impact site, spallation, and delamination are seen from the figure. Service life: '22'.

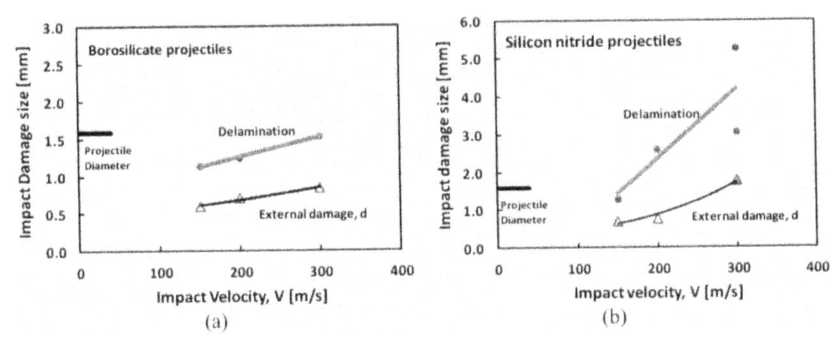

Figure 11. Plots of delamination crack and 'external' damage ('d') sizes as a function of impact velocity, determined in EB-PVD TBC airfoils impact by: (a)1.6-mm diameter borosilicate glass ball projectiles; (b) 1.6-mm diameter silicon nitride ball projectiles. The average data on the external damage ('d') from Figs. 3 and 6 were used for simplicity.

Comparison of Damage by Glass, Silicon Nitride, and Steel Projectiles

Figure 12 compares approximately external damage sizes ("d") determined by glass and silicon nitride projectiles from this work and by steel ball projectiles from the previous work [42]. For a given impact damage, the damage was greatest for steel projectiles, intermediate for silicon nitride projectiles, and least for glass projectiles. The reason for this different damage response may be ascribed to several factors such as projectile materials, degree of elastic/plastic deformation of projectiles as well as of coatings/substrate, and dynamic effects. Although not presented here, when the data was plotted as a function of impact kinetic energy ($=mV^2/2$ with m being the mass of a projectile), the data seemed to follow a general trend of linearity with kinetic energy; i.e., the greater kinetic energy, the greater impact damage, and vice versa. Hence, the degree of kinetic energy of individual projectiles for a given impact velocity would have the most significant effect on the severity of impact damage. However, it should be kept in mind that this is not always true. For example, when monolithic ceramics were impacted by steel and ceramic ball projectiles, ceramic ball projectiles always resulted in much greater damage than steel ball counterparts, due to generation of more significant Hertzian contact damage by the ceramic projectiles [15,45].

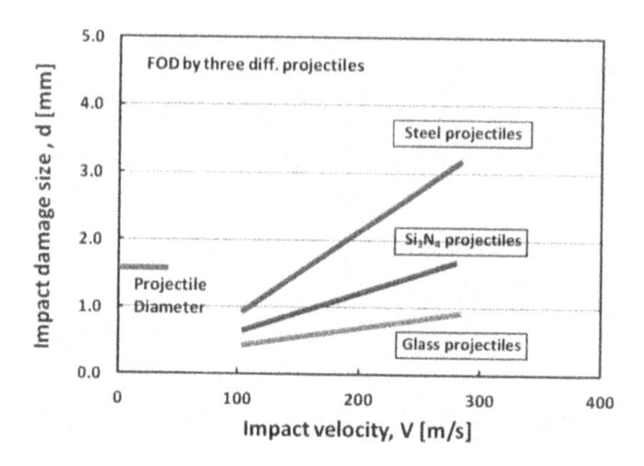

Figure 12. Overall comparison of external impact damage ('d') as a function of impact velocity with three different projectile materials determined from EB-PVD TBC airfoils impacted by 1.6-mm diameter ball projectiles. The silicon nitride and glass data were from this study, while the steel data were from the previous study [42].

Prediction of Impact Force

One may wonder how much impact force was acting upon impact by different projectile materials. Assuming that the impact event is quasi-static, a first-order approximation of impact force was made in this section using the energy balance principle in conjunction with contact yield pressure analysis that were applied to monolithic ceramics and CMCs [15,46]. Figure 13-a shows a relationship between contact indent area ($A=\pi d^2/4$) and indent force (P), determined via static indentation experiments using ball projectiles of three different materials. Independent of projectile materials, a linearity between A and P was reasonably established. The linearity implies that macroscopically the average 'contact yield pressure' ($=P/A$) would remain almost constant. The contact yield pressure, p_y, is defined as [15,46]

$$p_y = \frac{dP}{dA} \approx \frac{\Delta P}{\Delta A} \tag{1}$$

The value of p_y was estimated to be $p_y \approx 2540$ MPa from the data in Fig. 13-a. The energy balance in impact event may be written as follows:

$$U_k = U_{el} + U_{pl} + U_L + U_{k,bc} \tag{2}$$

where U_k is the kinetic energy of a projectile, U_{el} is the elastic deformation energy of projectile and TBCs/bondcoat/substrate, U_{pl} is the plastic energy of projectile and TBCs/bondcoat/substarte, U_L is th energy loss associated with friction and cracking such as spallation and delamination, and $U_{k,bc}$ is the bouncing-back kinetic energy of a projectile. Assuming the projectiles are more rigid than TBC systems, the energies may be expressed as follows:

$$U_k = mV^2/2$$
$$U_{pl} = \int F dz \qquad (3)$$
$$U_{k,bc} = mV_{bc}^2/2$$

where m is the mass of a projectile, V is the impact velocity, F is the impact force, V_{bc} is the bouncing-back velocity of a projectile. The impact depth z is defined in Fig. 13-b and is geometrically related to the impact impression diameter d and the projectile diameter D as

$$d = (4Dz - 4z^2)^{1/2} \qquad (4)$$

A relationship between impact velocity and bouncing-back velocity of a projectile is expressed

$$e = -V_{bc}/V \qquad (5)$$

where e is defined as a coefficient of restitution.

Assuming that the impacting kinetic energy is mainly consumed to plastically deform the TBCs/bondcoat/substrate with little energy loss either in elastic deformation (compared to the plastic component) or in cracking (delamination, spalling), one can obtain the following equation from Eqs. (2)-(4):

$$m(1-e^2)V^2/2 = \pi p_y (Dz^2/2 - z^3/3) \qquad (6)$$

The value of z in Eq. (6) can be solved as a function of impact velocity for the given values of p_y, m, and e. Once z is solved, the impact force F can be calculated using the following equation

$$F = [\pi(Dz - z^2)] p_y \qquad (7)$$

The resulting prediction of impact force is shown in Fig. 13-c where a value of e=0.3 was used. As seen in the figure, F increases almost linearly with increasing V. For a given V, the impact force is highest for steel projectiles, intermediate for silicon nitride projectiles, and lowest for glass projectiles. At V=300 m/s, the impact force reaches up to around 3.5, 2.3, 2.0 kN, respectively, for steel, silicon nitride, and glass projectiles. The degree of impact damage associated with different projectile materials as seen in Fig. 12 is thus related to the magnitude of impact force. It should be noted that the difference in F between e=0 to 0.5 was observed insignificant.

Since impact force measurements were not feasible due to the geometrical complexity of the airfoils, the impact-force prediction was not verifiable. Instead, the data on impact damage size (d) were used for verification because of their availability, although limited. The result is shown in Fig. 13-d, where the predicted d (solid lines) is compared with the experimental data for the three projectile materials. Despite limited data and some discrepancy, the overall prediction is in reasonable agreement with the experimental data. However, the approach made here to predict impact force was only of first-order approximation because of many simplifying assumptions and uncertainties associated with the violent and random nature of impact events. Pertinent test instrumentation/scheme as well as modeling should be sought to better assess impact force and related dynamics as well.

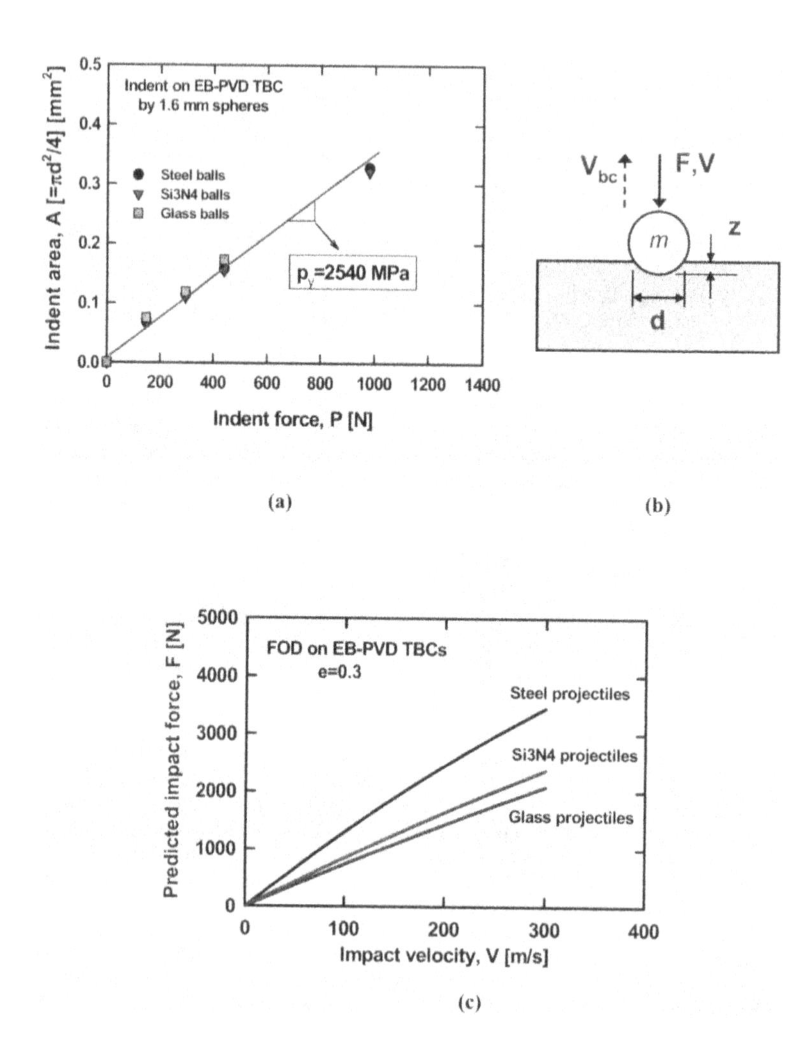

(a) (b)

(c)

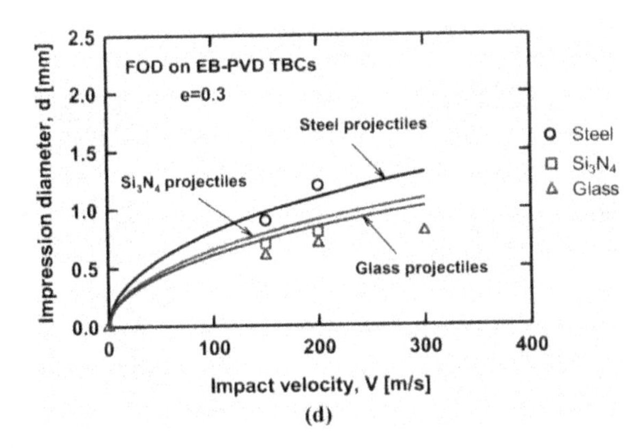

(d)

Figure 13. Prediction and verification of impact force: (a) Indent area (A) *vs.* indent force (P) in static indent experiments; (b) Deformation geometry of TBCs with regard to a spherical projectile; (c) Predicted impact force for three different projectiles materials; (d) Comparison of impression damage diameter (d) between prediction (solid lines) and experiment (symbols). Each data point in Fig.13-d represents the grand average of the data shown in Figs. 3 and 6 for glass and ceramic projectiles and from the previous study for steel projectiles [42].

Consideration Factors in FOD in TBCs

Exploration of FOD behavior of TBCs in airfoil components is an enormous task involving a lot of complicated variables. It is also associated with high degree of complexity due to dynamic effects or stress wave interactions. Multidisciplinary approaches in materials, service environment, mechanics, modeling, and fabrication/processing are needed to systematically understand impact behavior of TBCs in actual components. A schematic showing numerous affecting factors is illustrated in Fig. 14. Included in the figure are the factors such as TBCs, bondcoats, dynamic effects, substrate, environments, impact condition, and impacting object.

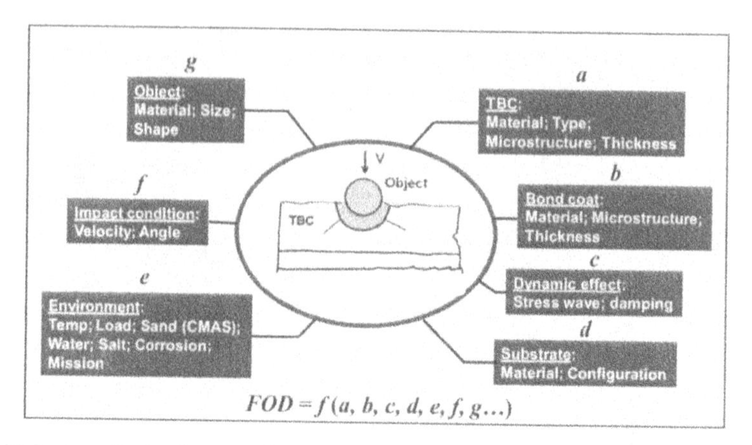

Figure 14. Factors affecting foreign object damage (FOD) in TBCs in airfoil components.

CONCLUSIONS

1) Impact damage of EB-PVD TBCs in airfoil components depended on both impact velocity and projectile material. For a given impact velocity, the impact damage was least for glass ball projectiles, intermediate for silicon nitride ceramic ball, and the greatest for steel ball projectiles. The effect of component service life between '3' and '22' was observed insignificant.
2) Delamination of TBCs along the topcoat/bondcoat interfaces was substantial in size typically two to three times greater than the externally observed damage.
3) Pertinent damage assessment tools or methods should be exercised to better characterize subsurface damage/morphology particularly for delamination cracking.
4) The first-order approximation of impact force based on energy balance principle was in reasonable agreement with experimental damage-size data.

Acknowledgements

This work was conducted through the support of the Office of Naval Research (ONR) and Dr. Dave Shifler.

REFERENCES

1. J. E. Ritter, S. R. Choi, K. Jakus, P. J. Whalen, R. G. Rateick, "Effect of Microstructure on the Erosion and Impact Damage of Sintered Silicon Nitride," *J. Mater. Sci.*, **26** 5543-5546 (1991).
2. Y. Akimune, Y. Katano, K. Matoba, "Spherical-Impact Damage and Strength Degradation in Silicon Nitrides for Automobile Turbocharger Rotors," *J. Am. Ceram. Soc.*, **72**[8] 1422-1428 (1989).
3. C. G. Knight, M. V. Swain, M. M. Chaudhri, "Impact of Small Steel Spheres on Glass Surfaces," *J. Mater. Sci.*, **12** 1573-1586 (1997).
4. A. M. Rajendran, J. L. Kroupa, "Impact Design Model for Ceramic Materials," *J. Appl. Phys*, **66**[8] 3560-3565 (1989).

5. L. N. Taylor, E. P. Chen, J. S. Kuszmaul, "Microcrack-Induced Damage Accumulation in Brittle Rock under Dynamic Loading," *Comp. Meth. Appl. Mech. Eng.*, **55** 301-320 (1986).
6. R. Mouginot, D. Maugis, "Fracture Indentation beneath Flat and Spherical Punches," *J. Mater. Sci.*, **20** 4354-4376 (1985).
7. A. G. Evans, T. R. Wilshaw, "Dynamic Solid Particle Damage in Brittle Materials: An Appraisal," *J. Mater. Sci.*, **12** 97-116 (1977).
8. B. M. Liaw, A. S. Kobayashi, A. G. Emery, "Theoretical Model of Impact Damage in Structural Ceramics," *J. Am. Ceram. Soc.*, **67** 544-548 (1984).
9. M. van Roode et al., "Ceramic Gas Turbine Materials Impact Evaluation," ASME Paper No. GT2002-30505 (2002).
10. D. W. Richerson, K. M. Johansen, "Ceramic Gas Turbine Engine Demonstration Program," Final Report, DARPA/Navy Contract N00024-76-C-5352, Garrett Report 21-4410 (1982).
11. G. L. Boyd, D. M. Kreiner, "AGT101/ATTAP Ceramic Technology Development," Proceeding of the Twenty-Fifth Automotive Technology Development Contractors' Coordination Meeting, p.101 (1987).
12. M. van Roode, W. D. Brentnall, K. O. Smith, B. Edwards, J. McClain, J. R. Price, "Ceramic Stationary Gas Turbine Development – Fourth Annual Summary," ASME Paper No. 97-GT-317 (1997).
13. (a) S. R. Choi, J. M. Pereira, L. A. Janosik, R. T. Bhatt, "Foreign Object Damage of Two Gas-Turbine Grade Silicon Nitrides at Ambient Temperature," *Ceram. Eng. Sci. Proc.*, **23**[3] 193-202 (2002); (b) S. R. Choi, J. M. Pereira, L. A. Janosik, R. T. Bhatt, "Foreign Object Damage in Flexure Bars of Two Gas-Turbine Grade Silicon Nitrides," *Mater. Sci. Eng.*, **A 379** 411-419 (2004).
14. S. R. Choi, J. M. Pereira, L. A. Janosik, R. T. Bhatt, "Foreign Object Damage of Two Gas-Turbine Grade Silicon Nitrides in a Thin Disk Configuration," ASME Paper No. GT2003-38544 (2003); (b) S. R. Choi, J. M. Pereira, L. A. Janosik, R. T. Bhatt, "Foreign Object Damage in Disks of Gas-Turbine-Grade Silicon Nitrides by Steel Ball Projectiles at Ambient Temperature," *J. Mater. Sci.*, **39** 6173-6182 (2004).
15. S. R. Choi, "Foreign Object Damage Behavior in a Silicon Nitride Ceramic by Spherical Projectiles of Steels and Brass," *Mat. Sci. Eng.*, **A497** 160-167 (2008).
16. S. R. Choi, "Foreign Object Damage Phenomenon by Steel Ball Projectiles in a SiC/SiC Ceramic Matrix Composite at Ambient and Elevated Temperatures," *J. Am. Ceram. Soc.*, **91**[9] 2963-2968 (2008).
17. R. T. Bhatt, S. R. Choi, L. M. Cosgriff, D. S. Fox, K. N. Lee, "Impact Resistance of Uncoated SiC/SiC Composites," *Mat. Sci. Eng.*, **A476** 20-28 (2008).
18. D. C. Phillips, N. Park, R. J. Lee, "The Impact Behavior of High Performance, Ceramic Matrix Fibre Composites," *Composites Sci.Tech.*, **37** 249-265 (1990).
19. A. R. Boccaccini, S. Atiq, D. N. Boccaccini, I. Dlouhy, C. Kaya, "Fracture Behavior of Mullite Fibre Reinforced-Mullite Matrix Composites under Quasi-Static and Ballistic Impact Loading," *Composites Sci. Tech.*, **65** 325-333 (2005).
20. Y. Leijiang, F. Ziyang, C. Qiyou, "Low Velocity Impact Damage Evaluation of 2D C/SiC Composite Material," *Advanced Materials Research*, **79-82** 1835-1838 (2009).
21. (a) S. R. Choi, D. J. Alexander, R. W. Kowalik, "Foreign Object Damage in an Oxide/Oxide Composite at Ambient Temperature," *J. Eng. Gas Turbines & Power, Transactions of the ASME*, Vol. **131**, 021301 (2009). (b) S. R. Choi, D. J. Alexander, D. C. Faucett, "Comparison in Foreign Object Damage between SiC/SiC and Oxide/Oxide Ceramic Matrix Composites," *Ceram. Eng. Sci. Proc.*, **30**[2] 177-188 (2009).
22. S. R. Choi, D. C. Faucett, and D. J. Alexander, "Foreign Object Damage in An N720/Alumina

Oxide/Oxide Ceramic Matrix Composite," *Ceram. Eng. Sci. Proc.*, **31**[2] 221-232 (2010).
23. K. Ogi, T. Okabe, M. Takahashi, S. Yashiro, A. Yoshimura, "Experimental Characterization of High-Speed Impact Damage Behavior in A Three-Dimensionally Woven SiC/SiC Composite," *Composites Part A*, **41**[4] 489-498(2010).
24. V. Herb, G. Couegnat, E. Martin, "Damage Assessment of Thin SiC/SiC Composite Plates Subjected to Quasi-Static Indentation Loading," *Composites Part A*, **41**[11] 1677-1685 (2010).
25. D. C. Faucett, S. R. Choi, "Foreign Object Damage in An N720/Alumina Oxide/Oxide Ceramic Matrix Composite under Tensile Loading," *Ceramic Transactions,* **225** 99-107, Eds. N. P. Bansal, J. P. Singh, J. Lamon, S. R. Choi (2011).
26. R. T. Bhatt, S. R. Choi, L. M. Cosgriff, D. S. Fox, K. N. Lee, "Impact Resistance of Evironmental Barrier Coated SiC/SiC Composites," *Mater. Sci. Eng.*, **A476** 8-19 (2008).
27. D. Hass, "Impact and Erosion Resistance of Novel Thermal/Environmental Barrier Coating Systems," presented at MS&T'11 Conference, October 16-20, 2011, Columbus, OH.
28. Advances in Ceramic Armor II-VII, *Ceram. Eng. Sci. Proc.*, **28-32** (2007-2011).
29. P. Lundberg, R. Renström, B. Lundberg, "Impact of Conical Tungsten Projectiles on Flat Silicon Carbide Targets: Transition from Interface Defeat to Penetration," *Int. J. Impact Eng.*, **32** 1842-1856 (2006).
30. V. S. Deshpande, A. G. Evans, "Inelastic Deformation and Energy Dissipation in Ceramics: A Mechanism-Based Constitutive Model," *J. Mech. Phys. Solids*, 56 3077-3100 (2008).
31. J. C. LaSalvia, J. W. McCauley, "Inelastic Deformation and Damage in Structural Ceramics Subjected to High-Velocity Impact," *Int. J. Appl. Ceram. Tech.*, **7**[5] 595-605 (2010).
32. S. R. Choi, D. Zhu, R. A. Miller, "Mechanical Properties/Database of Plasma-Sprayed ZrO_2-8wt% Y_2O_3 Thermal Barrier Coating," *Int. J. Appl. Ceram. Technol.*, **1**[4] 330-342 (2004).
33. R. W. Bruce, "Development of 1232°C (2250°F) Erosion and Impact Tests for Thermal Barrier Coatings," *Tribology Trans.*, **41**[4] 399-410 (1998).
34. M. Fathy, W. Tabakoff, "Computation and Plotting of Solid Particle Flow in Rotating Cascades," *Computers & Fluids*, **2**[1-A] 1-15 (1974).
35. X. Chen, M. Y. He, I. Spitsberg, N. A. Fleck, J. W. Hutchinson, A. G. Evans, "Mechanisms Governing the High Temperature Erosion of Thermal Barrier Coatings," *Wear*, **256** 735-746 (2004).
36. J. R. Nicholls, M. J. Deakin, D. S. Rickerby, "A Comparison Between the Erosion Behavior of Thermal Spray and EB-PVD Thermal Barrier Coatings," *Wear*, **233-235** 352-361 (1999).
37. X. Chen, R. Wang, N. Yao, A. G. Evans, J. W. Hutchinson, R. W. Bruce, "Foreign Object Damage in a Thermal Barrier Systems: Mechanism and Simulations," *Mater. Sci. Eng.*, **A352** 221-231 (2003).
38. A. G. Evans, N. A. Fleck, S. Faulhaber, N. Vermaak, M. Maloney, R. Darolia, "Scaling Laws Governing the Erosion and Impact Resistance of Thermal Barrier Coatings," *Wear*, **260** 886-894 (2006).
39. J. R. Nicholls, Y. Jaslier, D. S. Rickerby, "Erosion and Foreign Object Damage of Thermal Barrier Coatings," *Material Science Forum*, **251**[1-2] 935-948 (1997).
40. J. R. Nicholls, R. G. Wellman, "Erosion and Foreign Object Damage of Thermal Barrier Coatings," RTO-MP-AVT-109, 20-1_20-30 (2003); presented at the *RTO AVT Specialist Meetings on "The Control and Wear Military Platforms,"* June 7-9, 2003, Williamsburg, VA.
41. M. W. Crowell, et al., "Experimental and Numerical Simulations of Single Particle FOD-Like Impacts of HPT TBCs," in preparation (2011).
42. D. C. Faucett, J. Wright, M. Ayre, S. R. Choi, "Foreign Object Damage (FOD) in Thermal Barrier Coatings," presented at MS&T Conference, October 16-21, Columbus, OH; to be published in *Ceram. Trans.*, (2012).

43. D. B. Marshall, A. G. Evans, "Measurement of Adherence of Residually Stresses Thin Films by Indentation I. Mechanics of Interface Delamination," *J. Appl. Phys.*, **56**[10] 2632-2638 (1984).
44. A. G. Evans, J. W. Hutchinson, "On the Mechanics of Delamination and Spalling in Compressed Films," *Int. J. Solids Structures*, 20[5] 455-466 (1984).
45. S. R. Choi and Z. Rácz, "Target Size Effects on Foreign Object damage in Gas-Turbine Grade Silicon Nitrides by Ceramic Ball Projectiles," ASME Paper No. GT2010-23574 (2010).
46. D. C. Faucett, D. J. Alexander, S. R. Choi, "Static Contact Damage in an N720/Alumina Oxide Ceramic Matrix Composite with Reference to Foreign Object Damage," *Ceram. Eng. Sci. Proc.*, **31**[2] 233-244 (2010).

SOLID PARTICLE EROSION OF THERMAL SPRAY AND PHYSICAL VAPOUR DEPOSITION THERMAL BARRIER COATINGS

F. Cernuschi, C. Guardamagna, L. Lorenzoni, S. Capelli
RSE – Ricerca per il Sistema Energetico, Via Rubattino, 54, 20134 Milano, Italy
F. Bossi
Dipartimento di Chimica, Materiali e Ingegneria Chimica 'Giulio Natta',Politecnico di Milano, Via Mancinelli, 7 20131 Milano Italy.
R. Vaßen
Forschungszentrum Jülich GmbH, Institut für Energieforschung IEF-1, 52425 Jülich, Germany
K. von Niessen
Sulzer Metco AG, Rigackerstr.16, CH-5610, Wohlen, Switzerland

ABSTRACT

Thermal barrier coatings (TBC) are used to protect hot path components of gas turbines from hot combustion gases. Electron beam physical vapour deposition (EB-PVD) coatings have a columnar microstructure that guarantees high strain compliance and better solid particle erosion than PS TBCs. The main drawback of EB-PVD coating is the deposition cost that is higher than that of air plasma sprayed (APS) TBC. Nowadays segmented APS coatings and PS - PVD™ have been developed to improve solid particle erosion of plasma sprayed TBCs .In this perspective standard porous APS, segmented APS, EB-PVD and PS - PVD™ were tested at 700°C in a solid particle erosion jet tester, with EB-PVD and standard porous APS being the two reference systems.

Tests were performed at impingement angles of 30° and 90°, representative for particle impingement on trailing and leading edges of gas turbine blades and vanes, respectively. Microquartz and alumina were chosen as the erodent. To investigate the effect of grain size distribution, erosion rates when fine and coarse alumina powders have been used. Furthermore, after the end of the tests, the TBC microstructure was investigated using electron microscopy to characterise the failure mechanisms taking place in the TBC.

INTRODUCTION

Ceramic thermal barrier coatings (TBCs) are widely applied for protecting hot path components of gas turbines from combustion gases. The state-of the-art of TBC is represented by Yttrium oxide partially stabilized Zirconium (YPSZ) oxide (7-8 wt.% Y_2O_3+ZrO_2) deposited onto the components either by Air Plasma Spray (APS) or by Electron Beam Physical Vapour Deposition (EB-PVD) [1].

As pointed out from many authors, owing to the deposition process, typical APS TBC microstructure consists in ceramic lamellar shaped splats, vertical intra-lamellar cracks, horizontal inter-lamellar pores and globular voids [2, 3]. EB-PVD coatings have mostly columnar microstructure that guarantees higher strain compliance but lower thermal insulation compared to APS TBC [4 –7]. Depending on deposition parameters a more or less pronounced feathery structure can be observed within each column.

In the last years dense vertically cracked APS TBC have been developed to improve strain compliance and erosion rate of APS TBC keeping the deposition costs lower than EB-PVD coatings. [8]. When sufficiently cracked, the strain compliance of these coatings is significantly enhanced by vertical cracks penetrating most of the TBC thickness. At the same time, the quite dense microstructure between vertical cracks guarantees a higher resistance to solid particle erosion, compared to standard porous APS TBC.

Solid particle erosion (SPE) is reported as one of the TBC failure modes both in aero and land based applications. In the case of aero gas turbines sand and volcanic ashes are the main erodent source while in the case of land based gas turbines, particles escaped from filters, or produced either within compressor stages or in the combustion chamber are responsible for erosion. As pointed out by Tabakoff, solid particles do not move along the flow streamlines and thus they impact on components eroding the protective coatings from the base materials [9]. Although erosion can occur along the whole turbine, Tabakoff pointed out that the leading edge of first rotating stage blades is the area mostly affected by erosion, but also at suction and pressure sides close to the trailing edge erosion can occur, especially when coarse sand particles (≅100 – 150 μm) are present. In this last case, the solid particle impingement angles differ from 90° (typical at leading edge) usually ranging between 25° and 30°.

In this work, an experimental study focused on the effect of composition and the size distribution of erosive powder on the erosion rate for APS and PVD TBC has been carried out. Furthermore the effect of impingement angle and speed on erosion rates has been investigated as well.

EXPERIMENTAL

The jet tester

The solid particle erosion resistance of the TBC systems has been tested using a Jet tester consisting in:

i. The test chamber containing the nozzle, the sample holder and the system for the erosive particle speed measurement.
ii. Two furnaces for heating the carrier gas of the erosive particles.
iii. The erosive particle powder feeder.
iv. The cyclone for powder filtering and removal.
v. The control board.

Two coils inserted within the two furnaces allow the carrier gas (typically compressed air) to be heated up to the desired temperature (from RT up to 700°C).

The solid erosive particles (proportioned by a powder feeder) are injected within the main gas stream and, through the nozzle, they are allowed to impinge at the desired angle on the surface of the tested specimen. The erosive particle speed is directly related to the pressure of the carrier gas. Two small furnaces inside the testing chamber keep the sample and nozzle at the desired temperature, respectively. The exhausted powder is then removed and filtered by the cyclone. Erosive particle speed measurement is performed using the rotating double-disk method [10]. Typical erosive flow rates and erosion loads range from about 2 g min^{-1} up to 5 g min^{-1}. and 1 – 4 kg m^{-2} s^{-1}, respectively.

The test procedure consists in the following steps:

- fix all the experimental parameters;
- weigh the specimen;
- position the specimen and wait until the testing temperature has been reached;
- make a fixed amount of powder impinge on the specimen surface;
- weigh the specimen.

This procedure is repeated until either an established total mass of erosive has impacted on the specimen surface or the top layer is completely eroded. After the end of the test, for each specimen, the evaluation of the erosion rate, as the slope of the straight line best fitting the experimental data, is performed by plotting the eroded weight as a function of the impacted mass. The incidence angle between the erosive particles and the specimen surface is set up by mounting samples within sample holders having the desired inclination in respect to the impinging particle stream.

Samples

For this study the following four TBC systems have been considered: standard porous APS TBC, highly segmented APS (HS-APS), EB-PVD and Plasma Spray - PVD™ TBCs. The PS-PVD process is based on the ChamPro™ technology of Sulzer Metco which comprises all those thermal spray processes performing under a defined and controlled atmosphere like LPPS, VPS and LVPS [11]. The PS-PVD™ process operates at lower pressures down to 0.5-2 mbar. Even though the PS-PVD™ work pressure (~1 mbar) is still much higher than the one used in conventional PVD processes (~10-3 mbar), the combination of a high energy plasma gun operated at a low pressure environment enables a defined evaporation of the injected powder material. This allows one to produce a controlled deposition out of the vapour phase producing a coating with a columnar structure similar to EB-PVD. Figure 1 shows the microstructure of all the three TBC systems. A more detailed description of the four TBC systems complete of deposition details can be found elsewhere [12].

For impingement angles equal to 90° and to 30°, 4 mm thickness 25 mm disks and rectangular 40x15x4 mm samples have been considered, respectively.

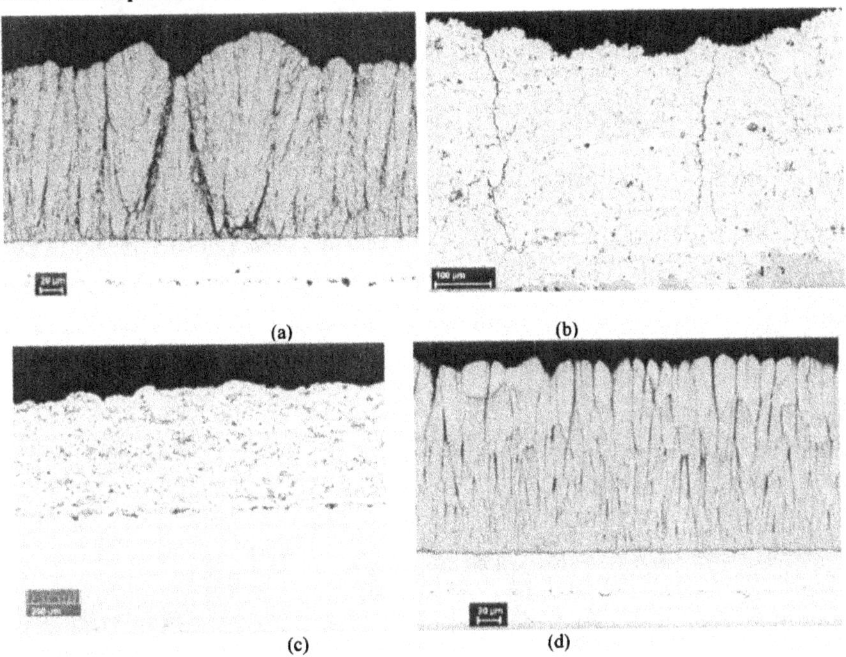

Figure 1 Scanning electron images (SEI) of a section of (a) PS-PVD™, (b) Highly segmented APS (c) standard porous APS and (d) EB-PVD TBC systems.

Testing conditions

Since in a previous work the solid particle erosion resistance of the same TBC systems was investigated by using as erosive microquartz (SiO_2 in the crystallographic hexagonal phase) powder (d10=68 μm, d50=122μm, d90=204 μm) and a certain set of testing conditions, for comparison purposes in this work, apart from the erosive powder, testing conditions have been chosen equal to those used in that previous work for as summarised in Table 1. In particular, testing temperature and the impingement speed have been set up equal to 700°C, and $v=104\pm5$ m s^{-1} (@30°) and $V=40\pm5$ m s^{-1} (@90°), respectively. As erodent, fine ((d10=10 μm, d50=16.7μm, d90=27.8 μm) and coarse (d10=70 μm, d50=126μm, d90=198 μm) α-alumina (Al_2O_3) powders have been used. Figure 2 shows the sharp edge morphologies and the grain size distribution of these two powders. The feed rate of erosive particles was set up to 2 g min^{-1}.

Limited to highly segmented APS TBC, the erosion rate has been estimated for several impingement speeds in the range 48 – 215 m s^{-1} to estimate the functional dependence of the erosion rate from the impingement speed.

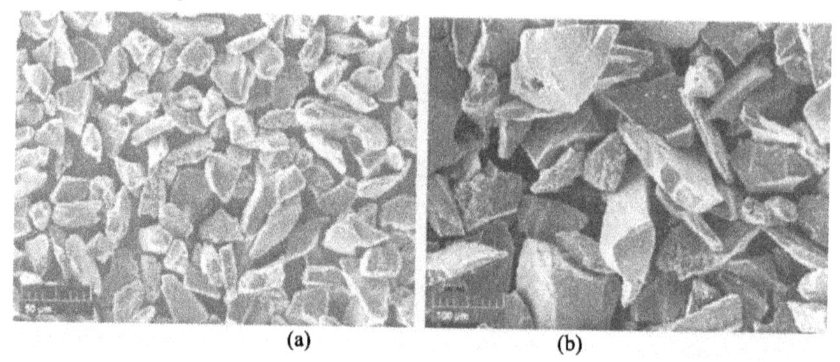

(a) (b)

Figure 2. Scanning secondary electron microscopy images of (a) fine and (b) coarse alumina powders used as erodent.

Table 1 Testing parameters

Parameter	**Value**
Testing temperature	700 °C
Impingement angles	30° and 90°
Erosive particles feeding rate	2 g/min
Maximum impingement speed @90°	40±5
Maximum impingement speed @30°	104±3

RESULTS AND DISCUSSION

Erosion rates

Figure 3 summarises the erosion rates for the three TBC systems when coarse alumina and microquartz powders are used (porous TBC was not tested in the previous work when microquartz was used). The main outcome is that erosion rates increase one order of magnitude when alumina powder is used instead of microquartz (with the same size distribution) for both impingement speeds and angles. The reasons for this significant increase are the higher hardness and density of the alumina powder compared to microquartz. Furthermore, at low speed and at 90° impingement angle, PS - PVD™ shows the best solid particle erosion resistance. At high speed and at 30° impingement speed, EBPVD or HS-APS performs better than PS-PVD coatings depending on the erodent.

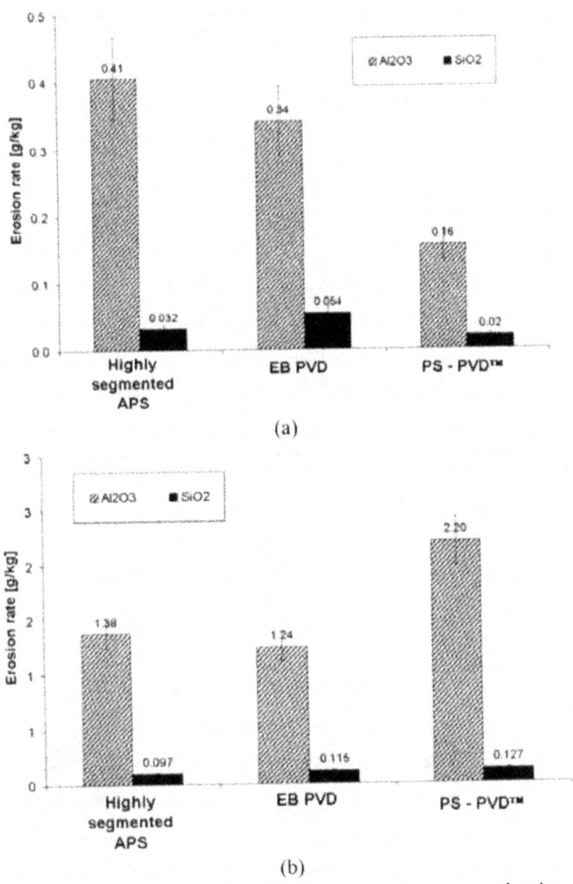

Figure 3. Solid particle erosion rates for three TBC systems when coarse alumina and microquartz powders are used. Testing speed and impingement angle (a) 40 m s^{-1} and 90° (b) 104 ms^{-1} and 30°, respectively.

Figure 4 summarises the effect of gain size distribution on the erosion rates for the four TBC systems. It is worth noting that at low impingement speed erosion rates decrease one order of magnitude or more going from coarse to fine powder. This is not true at higher impingement speed with impingement angle 30°, in fact a much less significant increase is observed. Also in the case of fine powder at low speed at 90° PS – PVD performs a little better than EB-PVD and vice versa at higher impingement speed and grazing angle.

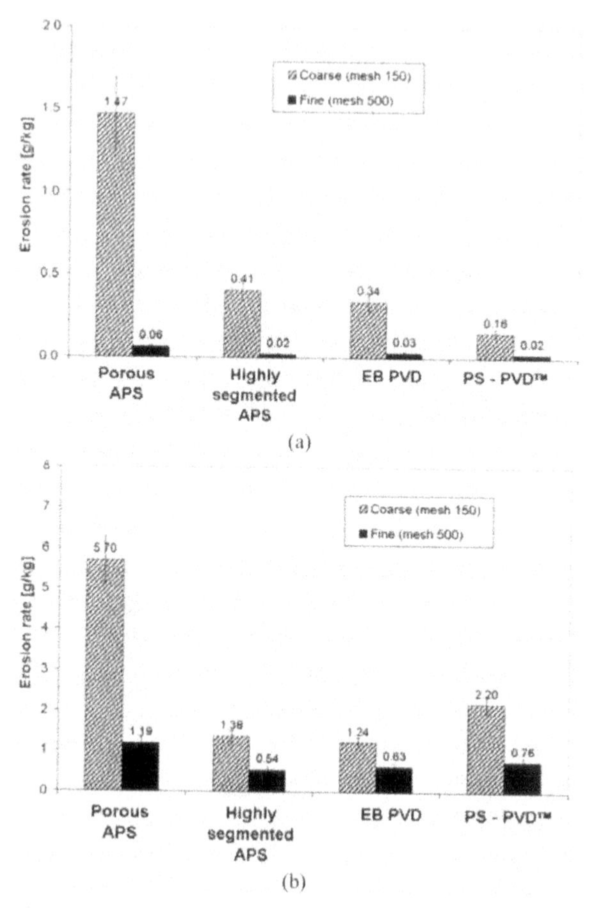

Figure 4. Solid particle erosion rates for the four TBC systems when fine and coarse alumina powders are used. Testing speed and impingement angle (a) 40 m s^{-1} and 90° (b) 104 ms^{-1} and 30°, respectively.

Figure 5 shows the erosion rate for HS-APS as a function of the impingement speed when coarse alumina powder and impingement angle equal to 30° have been set up. The fitting of the experimental data has been performed using a power law since it might be expected that the erosion rate w_e for brittle materials is proportional to the impingement speed component perpendicular to the target surface:

$$w_e \propto (v)^n \tag{1}$$

where v and n are the particle speed and an exponent (for bulk ceramic $n \cong 3$ [13]), respectively. In the present case, the value of the index n resulted a little bit higher than 2, as also previously observed by authors [12], although also values closer to 3 are reported in the literature [14]. It is worth noting that n values range between 1.7 and 2.8, when a confidence interval of 90% is considered.

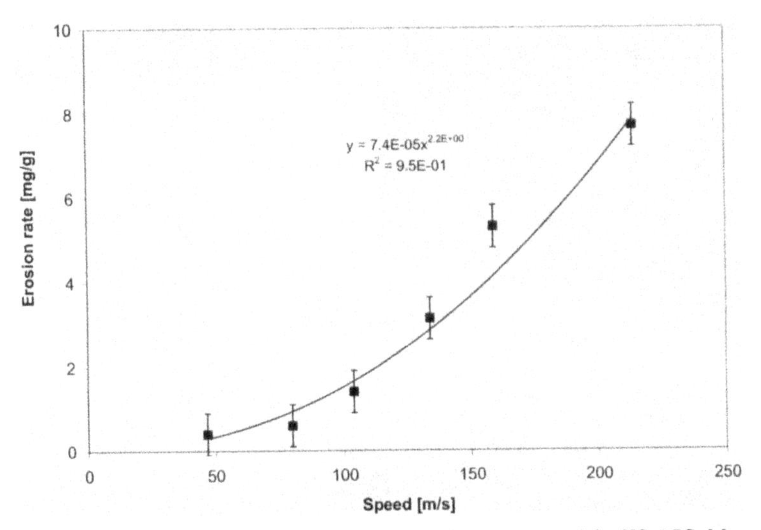

Figure 5. Solid particle erosion rate as a function of the impingement speed for HS-APS. Measurements have been performed using the impingement angle equal to 30°.

Comparing results with literature, when microquartz was used the erosion rates for all the samples resulted two orders of magnitude lower compared to those reported in the literature [15 – 20]. In particular, results in the literature refer to tests carried out using alumina erosive particles with typical impingement speeds and grain size in the range 140 – 360 m s^{-1} and 27 μm up to 200 μm, respectively. Considering the erosion rate for HS-APS when coarse alumina powder is used and an impingement speed in the same range (i.e. v=214 m s-1) erosion rates agree with those reported in the literature. This means that impingement speed, hardness and density of erodent particles can fully explain the previously observed difference of the erosion rates.

Failure mechanisms of the TBCs

SEM images taken along the section of tested samples (see Figure 6) confirm that for APS TBC the damage mechanism is related to multiple splat removal through crack propagation along the boundaries of cluster of splats as pointed out by Eaton and Novak [17]. In the case of Porous TBC a smoothing and densification of the upper layer can be observed as well. In the case of PS - PVD™ and EB-PVD samples, following the nomenclature of Wellman and Nicholls [15], mode II is the main damage mechanism. It consists in the formation of a densified thin layer on the top of the TBC caused by the multiple impacts of erodent particles (Figure 7). The progressive erosion of columns takes place by crack propagation at the interface between the densified layer and the as deposited columnar structure. The thickness of this upper dense layer increases with the impingement speed and the size of particles. Moreover, away from the maximum erosion area, for coarse powder at low speed also mode I damage mechanism is observed (see Figure 8). In this case, when particles impact on EB-PVD TBC surface with a sufficiently low speed, the response of the TBC is only elastic and cracking parallel to the surface are caused by tensile stresses promoted by the elastic waves propagating forward and backward along each single columns all around the impingement site.

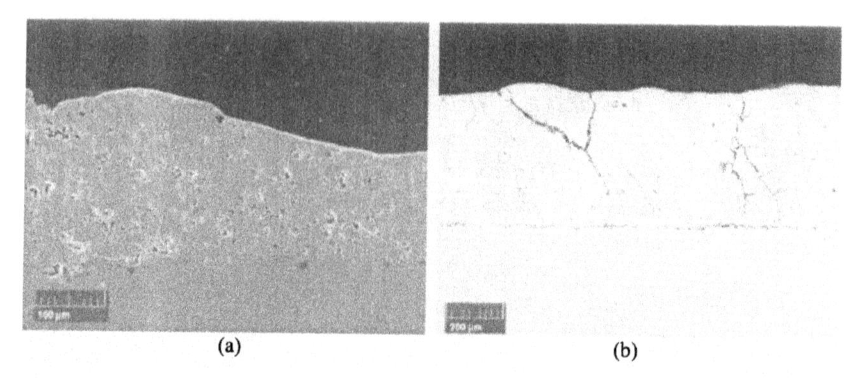

(a) (b)

Figure 6. SEM image of the eroded area for (a) standards porous and (b) HS – APS samples. A smooth and densified upper layer can be observed along the section of porous TBC. A crack joining along vertical cracks produced by solid particle impacts are clearly visible .

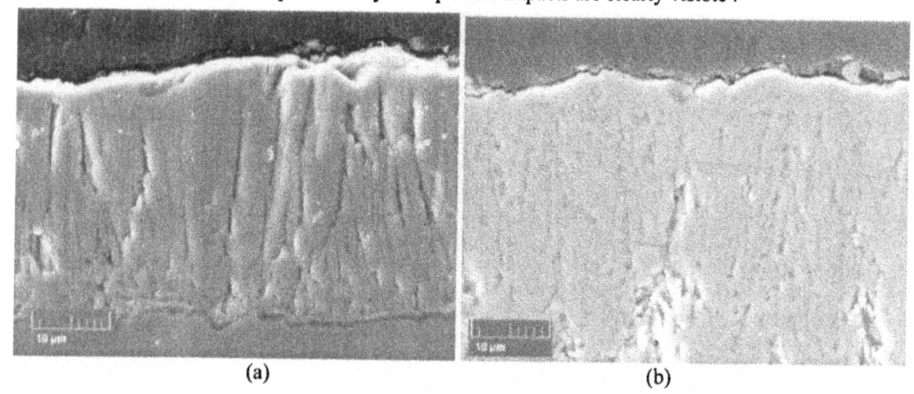

(a) (b)

Figure 7. SEM images of the eroded area for (a) EBPVD and (b) PS-PVD TBC. A thin densified layer promoted by the multiple impacts of particles on the top of the coatings can be distinguished (damage mechanism mode II).

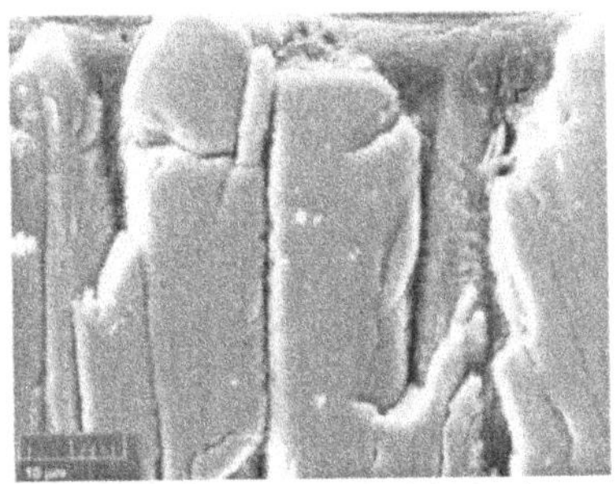

Figure 8. EBPVD sample tested at 40m s $^{-1}$. SEM image of the section away from the maximum erosion area. Cracking of individual columns due to solid particle impact are visible (Mode I damage mechanism).

CONCLUSIONS

Four different TBC systems were tested at 700°C in a solid particle erosion jet tester. Fine and coarse alumina powders representative of size distribution of sand and fly ashes, respectively, were used and erosion rates were compared to those reported in a previous work where microquartz was used as erodent. The impingement speed has been set up equal to 40 m s^{-1} and 104 m s^{-1} for impingement angle equal to 90° and 30°, respectively.

The main results can be summarised as follows:

- The PS - PVD™ coatings showed the lowest erosion rate at low speed and 90° impingement angle while at higher speed and 30° impingement angle EBPVD coatings perform better than all the other TBC systems.
- EB-PVD and the highly segmented APS coatings showed comparable erosion rates.
- Standard porous APS coating shows a significantly poor erosion resistance.
- Erosion rates increase more than one order of magnitude when the hardness and density of erodent increase from microquartz to alumina.
- Erosion rates increases when the size distribution of erodent is increased from mesh 500 to 150.
- The erosion rate vs. the impact speed seems to show a quadratic-cubic dependence
- In APS coatings the damage mechanism is related to multiple splats removal through crack propagation along splat boundaries.
- For PS - PVD™ and EB-PVD TBC a densification of a thin layer on the top of the TBC occurs and the erosion takes place by crack propagation at the interface between densified layer and the underneath TBC.

ACKNOWLEDGEMENTS

This work has been partially financed by the Research Fund for the Italian Electrical System under the Contract Agreement between RSE (formerly known as ERSE) and the Ministry of Economic Development - General Directorate for Nuclear Energy, Renewable Energy and Energy Efficiency stipulated on July 29, 2009 in compliance with the Decree of March 19, 2009.

REFERENCES

1. W. Beele, G. Marijnissen, A. Van Lieshout, The Evolution of Thermal Barrier Coatings: Status and Upcoming Solutions for Today's Key Issues, *Surf. Coat. Technol.*, 120-121 (1999), pp. 61-67.

2. J.A. Thompson, T.W. Clyne, 'The Effect of Heat Treatment on the Stiffness of Zirconia Top Coats in Plasma-Sprayed TBCs, *Acta mater.* 49, (2001), 1565.
3. S. Leigh, CC Berndt., Quantitative Evaluation Of Void Distributions Within A Plasma Sprayed Ceramic , *J Am Ceram Soc* 1999, 82 (1), 17.
4. R. A. Miller, "Thermal Barrier Coatings for Aircraft Engines: History and Directions," *J. Therm. Spray Technol.*, 6 [1] 35–42 (1997).
5. S. Bose and J. DeMasi-Marcin, "Thermal Barrier Coating Experience in Gas Turbine Engines at Pratt & Whitney," *J. Therm. Spray Technol.*, 6 [1] 99–104 (1997).
6. Z. Mutasim and W. Brentnall, "Thermal Barrier Coatings for Industrial GasTurbine Application: An Industrial Note," *J. Therm. Spray Technol.*, 6 [1] 105–108 (1997).
7. P. Scardi, M. Leoni, F. Cernuschi, A. Figari, Microstructure and heat transfer phenomena in ceramic Thermal Barrier Coatings, *J. Am. Ceram Soc.* 84[4], 827 – 835, (2001).
8. R. Vaßen, M. Ahrens, A.F. Waheed, D. Stöver, The Influence of the Microstructure of Thermal Barrier Coatings Systems on Sintering and Other Properties, in: E.L. Lugscheider, C.C. Berndt (Ed.), Tagungsband Conference Proceedings, Proceedings of International Thermal Spray Conference, Essen, Germany, 2002, DVS, Germany, 2002, pp. 879–883.
9. Tabakoff W. Investigation of coatings at high temperature for use in turbomachinery. *Surf Coat Technol* 1989;30/40:97–115.
10. ASTM G76 - 95 Standard Test Method for "Conducting Erosion Test by Solid Particle Impingement Using Gas Jets".
11. P. Ambühl, P. Meyer, Thermal Coating Technology in Controlled Atmospheres (ChamProTM), Proceedings of the ITSC, E. Lugscheider, P.A. Kammer, 1999 (Düsseldorf, Germany), DVS-Verlag, 1999, pp. 291-92.
12. F. Cernuschi, C. Guardamagna, L. Lorenzoni, S, Capelli, M. Karger, R. Vassen, K. Von Niessen, N. Markoscan, J. Menuey, C. Giolli, Solid particle erosion of thermal spray and physical vapour deposition thermal barrier coatings, *Wear* 271 (2011) 2909– 2918
13. A.W. Ruff, S.M Widerhorn "Erosion by Solid Particle Impact" in *Treatise on Materials Science and technology* Vol. 16 Erosion, C.M. Preece Ed., Academic Press 1979.
14. D. Zhu, R.A. Miller, M.A. Kuczmarski, Development and life prediction of erosion resistant turbine low conductivity thermal barrier coatings, NASA report TM-2010-215669.
15. R. G. Wellman, J.R. Nicholls, A review of the erosion of thermal barrier coatings, *J. Phys D: Appl. Phys.* 40 (2007) R293 – R305.
16. R. G. Wellman, J.R. Nicholls, K. Murphy, Effect of microstructure and temperature on the erosion rates and mechanisms of modified EB-PVD TBCs, *Wear*, 267, (2009), 1927 – 1934.
17. H.E. Eaton, R.C. Novak, Particulate erosion of plasma-sprayed porous ceramics, *Surf. Coat Technol.*, 30 (1987), 41 – 50
18. W. Tabakoff Temperature erosion resistance of coatings for use in gas turbine engines, *Surf. Coat. Technol.* 52, (1992) 65 – 79.
19. W. Tabakoff, V, Shanov, Erosion rate testing at high temperature for turbomachinery use, *Surf. Coat. Technol.* 76 – 77 (1995) 75 – 80.
20. B.Z. Janos, E. Lugscheider, P. Remer, effects of thermal ageing on the erosion resistance of air plasma sprayed zirconia thermal barrier coatings, *Surf. Coat Technol.*, 113 (1999), 278 – 285

EFFECT OF HARD THIN FILM COATINGS ON TRIBOCHEMICAL FILM BEHAVIOR UNDER LUBRICATED SLIDING CONTACT

C. Lorenzo-Martin, O. O. Ajayi, S. Torrel, N. Demas and G.R. Fenske.
Energy Systems Division
Argonne National Laboratory
Argonne, IL 60439

ABSTRACT

Most of tribological components such as gears and bearings which are engaged in sliding or rolling contacts are made of ferrous materials. In order to reduce friction and wear, lubricants are usually applied to components. The lubricants are normally formulated with functional additives such as anti-wear, friction modifier, anti-oxidant, etc. These additives react with the ferrous materials surface to form tribochemical surface films. A variety of hard thin-film coatings are increasingly being used in tribological components. These coatings can enhance the friction and wear behavior of the components. The impacts of thin film coatings on surface reactions by lubricant additives, and hence their effectiveness in reducing friction and wear is currently unclear. In this study, we evaluated the friction and wear behavior of several commercially available thin-film ceramic coatings when lubricated with unformulated and fully formulated synthetic oils. In tests with unformulated lubricant, friction coefficient behavior is similar to the uncoated steel. While in tests with fully formulated lubricants, a variety of behaviors were observed for different coatings. These behaviors are explained in terms of the differences in tribochemical film formation and properties on the coatings surfaces.

INTRODUCTION

Tribological components in machinery, such as gears, bearings etc., are usually lubricated either with oil or grease. Most of these components operate under severe sliding or rolling contact conditions and adequate lubrication is essential to optimal operation and prevention of failure. In recent years, lubricant formulation has become a science and a multi-millionaire a year business. The new goal of lubricants goes far beyond reliability issues. Indeed, much effort and financial resources are being dedicated to the engineering of smart lubricant to optimized tribological properties under very specific applications. Energy saving by reducing friction and minimal environmental impact are two important considerations in new lubricants design.

Lubricants used in various tribological applications are usually formulated with additives for optimal performance[1,2]. The commercial oil lubricants consist primarily of a basestock and an additives package. The basestock can be mineral-oil based or synthetic-oil (such as poly-alpha-olefin (PAO), ester or polyol) based. Additives normally accounts for 4-10% of the total weight in fully formulated lubricants. Most of the efforts in term of lubricant formulation are devoted to producing the right combination of basestock and additives with a particular or desirable functionality. There are many different types of additives with specific functions. The additives can be divided into four categories with regards to functions: (1) Additives for lubricity; such as antiwear to reduce wear, friction modifier for friction control and extreme pressure to prevent catastrophic failure. (2) Additives to control chemical breakdown; such as antioxidant to retard the degradation of the oil by air via oxidation, corrosion inhibitor to retard oxidation of the metallic surfaces in contact, and detergent to clean and neutralize oil impurities. (3) Additives for contaminant control; such as dispersants to keep contaminants suspended in the oil and prevent them from coagulating. (4) Additives for viscosity control: such as viscosity modifiers to keep viscosity higher at elevated temperatures.

In lubricated tribological contacts, different regimes of lubrications occur depending on the severity of the contact and the properties of the lubricant. In hydrodynamic regime the surfaces of the two components in contact are completely separated by lubricant fluid film. In elasto-hydrodynamic

(EHD)/ mixed regime there is some limited interaction between the surface materials of the tribological contacts. The interactions are mostly elastic in this regime, hence it is called EHD. Boundary regime is characterized by high severity contacts and extensive interaction between the surface materials, the so-called metal-to-metal contact. In this regime, friction and wear are higher than in other regimes and the lubricant additives play an important role. Additives in the lubricant interact with the surface material to produce a tribochemical film also known as boundary film. Depending on the nature of the additives, the resulting film can have different properties (e.g. low friction, wear resistant, etc...). Under hydrodynamic and EHD lubrication regimes, friction and wear are mostly determined by the lubricant fluid film properties. Under boundary lubrication regime, friction and wear are determined by the simultaneous shear of the fluid film, boundary film and the near surface material.

Existing lubricants additives are usually formulated for ferrous materials because most of the tribological components are made of steel or cast iron. However, over the last decade or so, thin film ceramic coatings are increasingly being used to enhance the performance of tribological components [3-5]. Most often they are used to solve persistent tribological problems or failures occurring with typical components.

Currently, there is very limited knowledge about the role of lubricant additives in tribological performance of ceramic coated surfaces. Since the ceramic coatings are relatively thin, they are usually applied to existing components with minimal changes to the system, including lubrication. The question of the nature of interaction between lubricants and thin film coatings on tribological performance has to be addressed. The interaction between the two technologies could be synergistic, antagonistic or no effect at all. In order to effectively integrate these two technologies (i.e. lubricant additives and coatings), adequate understanding of the interaction between them is needed. Hence the goal of the present study is to assess the impact of thin-film ceramic coatings on the formation and performance of tribochemical boundary films.

EXPERIMENTAL DETAILS

Four commercially available thin film ceramic coatings namely: titanium nitride (TiN), titanium carbon nitride (TiCN), titanium aluminum nitride (TiAlN), and metal-doped amorphous (a-C:H:Me + a-C:H) diamond like carbon coating (Me-DLC) were used for the present study. The performance of the coatings was compared to baseline uncoated case-carburized and hardened 4120 steel material, which is also the substrate for the coatings. Thin-film hard ceramic were deposited on steel flats of dimensions 38mm x 5.1mm x 6.3mm by physical vapor deposition technique to a thickness ranging between 1-5 microns. Some of the coatings relevant properties; thickness, hardness and roughness are summarized in table I. Uncoated case carburized steel flat substrate has a hardness 60 R_c and surface roughness of 19 nm R_a.

Oil-lubricated friction and wear tests were conducted with a roller-on-flat contact configuration in reciprocating sliding as shown in Figure 1 (schematically and picture). The counterface roller is made of heat hardened 4140 steel (Ø 6.3 mm, 10mm length) with surface roughness of 200 nm R_a. Tests were conducted with three different synthetic lubricants: basestock poly-alpha-olefin (PAO10) without additives and two fully formulated commercially available lubricants. One of the lubricants (lubricant A) contained additive package that was optimized for friction control, while the other lubricant (lubricant B) has additives package optimized for wear protection. Some of the lubricants properties as provided by the manufacturer are summarized in table II. Tests were conducted at 150N normal load (which imposes a 0.33 GPa Hertzian contact pressure), 0.5 Hz reciprocating frequency (30 rpm.), 10mm stroke length (equivalent average linear speed 10mm/s), 100 °C temperature and for a duration of 180 minutes. Friction coefficient was continuously measured during the testing. Wear on flats and the rollers were evaluated at conclusion of the test. Worn surfaces were characterized by optical and scanning electron microscopy.

Table I: Properties of thin-film hard ceramic coatings

Coatings	Deposition method	Type	Thickness (µm)	Manufacture Hardness (Hv)	Nano Hardness (Gpa)	Roughness (nm)
TiN	PVD	Monolayer	3.7	2900	29	87
TiCN	PVD	Monolayer	0.9	4000	39	49
AlTiN	PVD	bilayer	2.4	3500	34.5	19
Me-DLC	PVD	multilayer	3.5	1500-3000	14.7	56

Table II: Properties of the lubricants

LUBRICANT	PAO	Synthetic A	Synthetic B
Viscosity 40°C (cSt)	71.1	233.5	132
Viscosity 100°C (cSt)	10.70	18.7	17.5
Viscosity Index	-	92	146
Flash point (°C)	272	235	221
Pour point (°C)	-51	-15, 5	-45
Density (kg/m³) at 15.6C	837	905	860

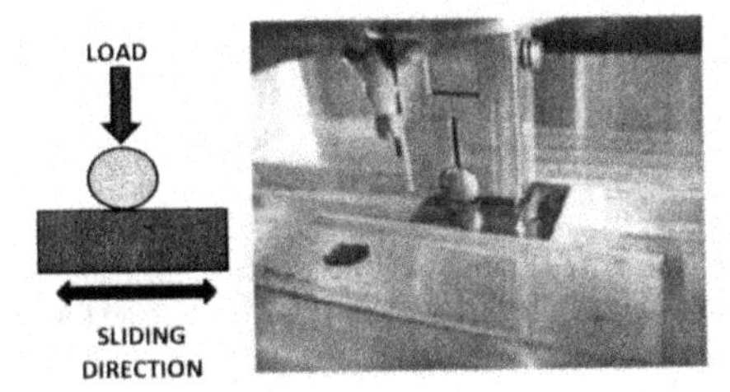

Figure 1: Roller-on-flat configuration for friction and wear testing under reciprocating sliding.

RESULTS

The friction behavior for the different coatings and the uncoated substrate when tested with basestock PAO without additives is summarized in figure 2. The friction coefficients varies with time for some of the materials tested (TiN, TiCN, and AlTiN coatings) and are nearly constant for the duration of test for other materials (uncoated substrate and the Me-DLC coating). Most of the variation in the friction at the early part of the tests can be attributed to the run-in process. After this initial period that can take 30-60 minutes, the frictional response for most of the materials reach a steady state with values typical for boundary lubrication regime (μ~0.13-0.15). AlTiN coating showed the largest decrease on its friction coefficient, during the run-in process; starting with a value of μ~0.3 the friction decreased to 0.18 in the first hour of testing and finally reached a value of μ~0.15 after 3 hours of testing. It should be noted also that the friction coefficient is highest for the AlTiN coating for the duration of the tests. TiN coating also showed considerable variation in friction with time. Although the friction coefficient at the beginning (μ~0.19) and to the end of the test (μ~0.13) didn't differ much, there was a significant reduction in the early stage of test, followed by an increase until steady state was finally reached towards the end of the test. The variation in frictional response for TiCN coating was somewhat similar to TiN, but lower in magnitude and much less noisy. Starting with a value of μ~0.17, the friction coefficient fell to 0.13 before increased and reached a steady value of μ~0.15. For Me-DLC coating the friction behavior was constant for the entire duration of the test with the lowest value of friction (μ~0.1). Also the uncoated substrate displayed an almost constant frictional response with a friction coefficient of μ~0.15 except for the very first minutes of run-in.

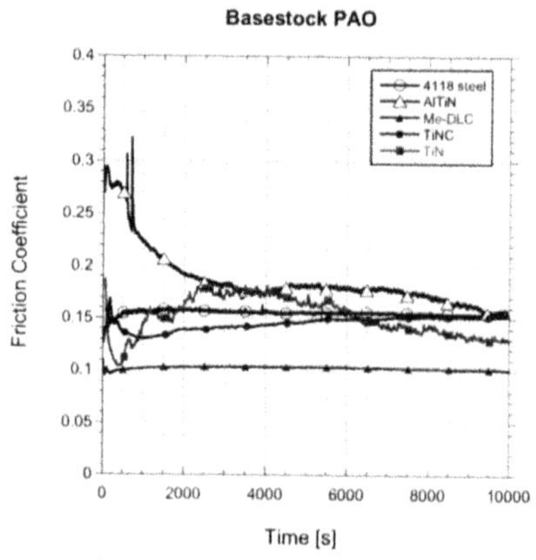

Figure 2: Friction coefficient variation with time for the different coatings and uncoated steel substrate lubricated with basestock (PAO10).

Results of the frictional behavior of different coatings and uncoated steel tested with fully formulated synthetic lubricant A are shown in figure 3. When tested in this lubricant, all materials showed a reduction in friction within the first 5-30 minutes of the test and reached a steady state friction coefficient. Because the friction reduction is much more than what was observed in the test with the basestock lubricant, the friction behavior in lubricant A is attributed to combined effect of run-in and interaction of the lubricant additives with the surface materials to form tribochemical films. With the exception of AlTiN, the steady state friction coefficient with values between 0.04-0.065 is considerably lower than typical boundary lubrication regime friction (μ~0.1-0.15). In ferrous materials, such low friction value under boundary lubrication regime is normally indicative of the formation of a low friction boundary films. Although the largest reduction in friction was observed for the uncoated steel substrate (from μ~0.125 to μ~0.04), some of the coatings also experienced a significant decreased in friction in the presence of this formulated oil: Me-DLC (from μ~0.1 to μ~0.05); TiN (from μ~0.2 to μ~0.055) and TiCN (from μ~0.15 to μ~0.065). Less friction reduction was observed in test with the AlTiN coating. After an initial decrease (from μ~0.2 to μ~0.1), very early in the test, friction coefficient became noisy and showed a somewhat gradual increase for the rest of the test. Although, lower friction coefficient was observed in all the tests with lubricant A (which was optimized for friction control) when compared to tests with the PAO basestock, the presence of coatings seems to reduce the effectiveness of the additives as the uncoated baseline steel exhibited the lowest friction coefficient of the materials tested.

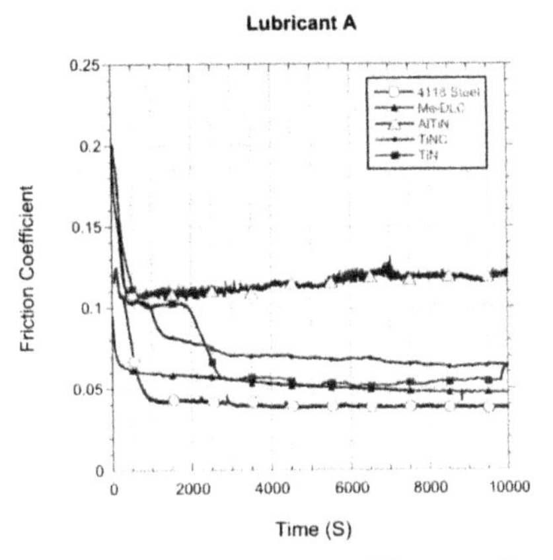

Figure 3: Friction coefficient variation with time for the different coatings and uncoated steel substrate with lubricant A.

Figure 4 shows the frictional behavior for different coatings and uncoated steel substrate tested with fully formulated synthetic lubricant B. Friction showed only a slight decrease after a run-in (10-15 minutes) and similar friction behavior was observed for all the coatings and uncoated steel substrate after the run-in period. A steady state friction coefficient was reached for all materials (μ~0.09-0.10). Again, the only exception to this frictional behavior is the AlTiN coating in which the test started with a friction coefficient that was substantially higher (about 0.2) than the remaining coating. There was a gradual decrease in friction coefficient until a steady state value (μ~0.115) was reached half the way into the test. Except for AlTiN, the coating material appears to have no effect on friction in tests with this lubricant. Other than the time needed to reach the steady state value, all the coatings and the uncoated steel substrate tested showed a similar steady state friction coefficient (μ~0.09-0.12).

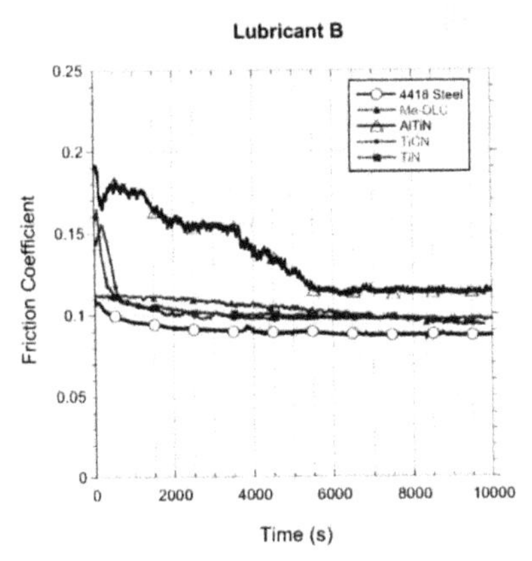

Figure 4: Friction coefficient variation with time for the different coatings and uncoated steel substrate with lubricant B.

At the conclusion of each test, the wear tracks of both the steel roller counterface and the flat surfaces were characterized by optical, 3D interferometry and scanning electron microcopy. Because most of the wear was produced in the softer counterface (roller), and wear on the flats was minimal (not measurable), an assessment of the wear was conducted by measuring the roller track width for each test. A summary of these measurements are documented in Figure 5. It is clear that wear in rollers tested against most of the coatings was higher than wear in roller tested against the uncoated baseline steel. Wear in roller tested against Me-DLC coating was comparable to roller wear against uncoated baseline steel. Formulated lubricants were more effective than basestock (PAO) in reducing roller wear when rubbed against uncoated baseline steel, especially in lubricant B (optimized for wear protection). Wear produced on the roller by the different coating seems to be independent of the type of lubricant, weather formulated or not. Except for Me-DLC, all other coatings are detrimental to counterface steel roller wear.

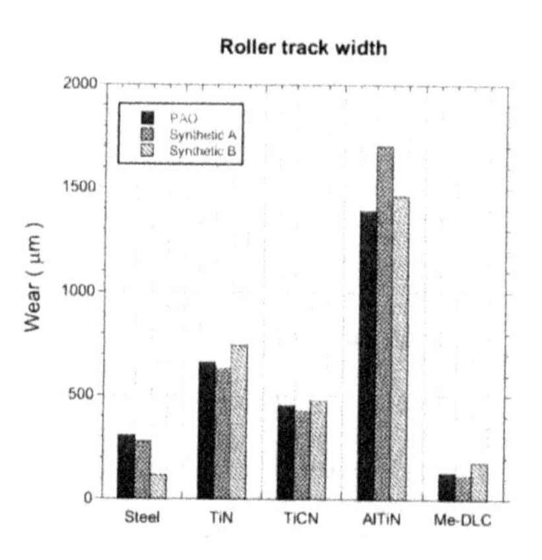

Figure 5: Roller track width after sliding against steel substrate and different thin-film coatings.

DISCUSSION

In friction and wear tests with basestock PAO that contains no additives, friction behavior for the uncoated steel substrate and all coatings are similar to one another and the measured friction coefficient values are typical for boundary lubrication regime. Because there are no additives in the lubricant, there was no formation of boundary film, and hence little or no effect on frictional performance. Wear measured in the steel roller counterface is higher when sliding against all coatings compared to the uncoated steel flat specimen; except for the DLC coating. The higher hardness of the coatings is responsible for more wear in the relatively softer steel counterface (roller). In the peculiar case of DLC coating, this material is known to protect the counterface surfaces under dry condition. This could be due in part to the transfer of a carbon layer from the DLC coating unto the counterface body, as has been reported in other studies under dry lubrication [6, 7]. Although the current study is lubricated, DLC also reduced wear in the counter-face as indicated in Figure 5. Evidence of a possible transfer of surface film from the coating into the roller has been observed (Figure 6). The mechanism by which this transfer takes place is not fully understood yet.

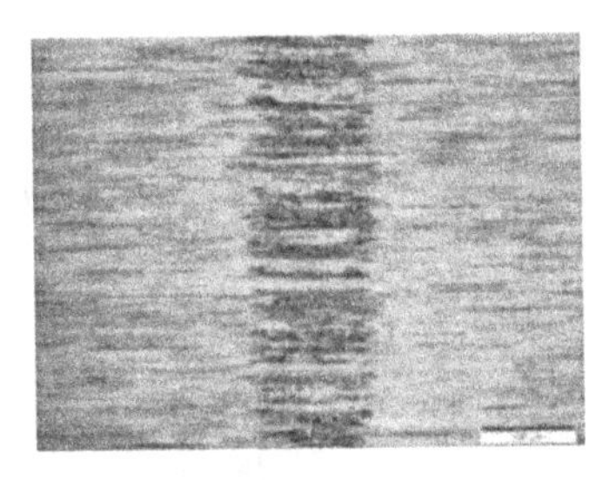

Figure 6: Carbon transfer film on steel roller slid against Me-DLC coating in PAO.

With fully formulated lubricant A, optimized for friction control, different friction behaviors were observed for the different materials tested. For the uncoated substrate (baseline) the friction coefficient started at 0.12 and decreased to a value of 0.04 due to the formation of a tribochemical film as shown in Figure 7A. This film fully covers the contact area and is responsible for the sustained and constant low friction behavior. For TiN and TiCN coatings, the friction trend is similar to the baseline, with a significant reduction in friction with time and reaching values for the friction coefficient of 0.055 and 0.065 respectively. There is evidence of limited localized tribochemical film formation and some metal transfer from the steel roller unto the coating surface, both present in the same areas, indicative of higher severity of contact in the localized areas (Figure 7B, 7C). For Me-DLC coating, friction started at a lower value (0.095) and decreased to a steady state friction coefficient value of (0.05). There is minimal film formation on the coating (figure 7D) but some evidence of film formation on the roller. It is unclear if this film is transfer carbon film or tribochemical reaction film. For AlTiN coating, friction coefficient started at a much higher value (0.2) and remained high until the end of the test (0.12). Extensive metal transfer from the roller (softer counterface) onto the coating was observed (Figure 7E).

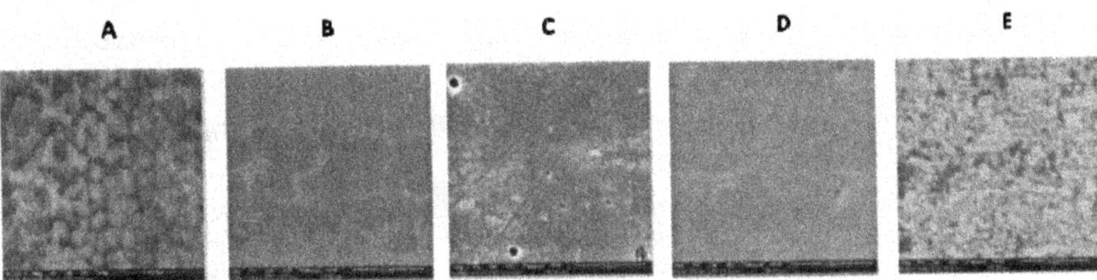

Figure 7: Micrographs of flat specimen surfaces tested with lubricant A; A): steel substrate, B) TiN, C) TiCN, D) Me-DLC and E) AlTiN.

For fully formulated lubricant B, optimized to minimize wear, similar friction behavior and values were observed for all coated and uncoated samples. For the uncoated steel substrate (baseline) the original surface topography was preserved (with no wear) and there was minimal tribochemical film formation; hence the friction remained constant (Figure 8A). TiN, TiCN and AlTiN coatings showed similar friction behavior and different degree of metal transfer and tribochemical film formation. They all produced more wear on the counterface steel roller but minimal to no wear in the coating themselves (figure 8B, 8C and 8D, respectively). For DLC coating, only slight decrease in friction with time was observed. There was no evidence of film formation on the coating, but evidence of some films was present on the roller wear track (Figure 8E). Recall that with unformulated PAO, minimal wear was observed in the roller because of the carbon transfer to protect counterface. With fully formulated lubricants wear is observed in the roller. This may be due to the additives interfering with carbon transfer to counterface when sliding against DLC coating, thereby inhibiting the wear protection provided by the carbon transfer from the coating.

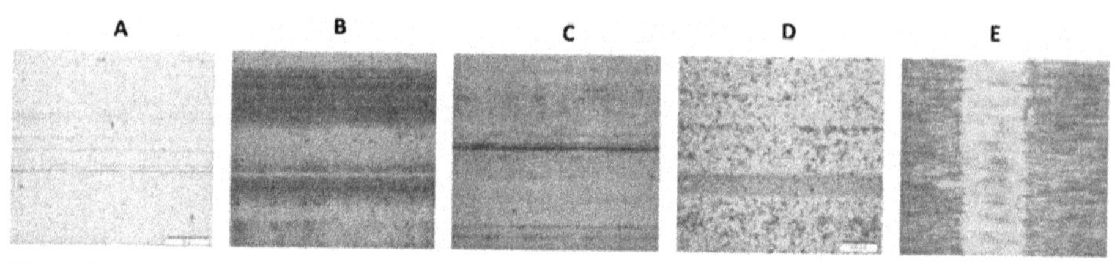

Figure 8: Micrographs of flat specimen surfaces tested with lubricant B; A): steel substrate, B) TiN, C) TiCN, D) AlTiN and E) steel roller tested against Me-DLC.

SUMMARY

Based on the experimental results from the present study, lubricant formulation with additives was effective in reducing friction and/or wear in steel-on-steel lubricated contact through the formation of surface tribochemical films as expected. The presence of thin-film ceramics coatings on one of the sliding surfaces sometimes has no effect and sometimes is detrimental to the effectiveness of tribochemical films produced from lubricated additives.

Me-DLC coating has beneficial effect by reducing both friction and wear in contacts lubricated with un-formulated basestock oil. In fully formulated lubricants, the DLC was not as effective, although still beneficial.

TiN and TiCN coatings both have minimal effect on friction behavior of formulated and unformulated lubricants, but produced more wear on the counterface. Evidence of possible tribofilm formation was observed.

AlTiN coating showed higher friction and wear of counterface in all the lubricants tested. There is no evidence of tribofilm formation on the coating.

Much more research work is needed to develop adequate understanding of the interactions between lubricant additives and thin-film coatings. This will facilitate the development and design of robust tribological components that can effectively integrate lubricant and coating technologies for optimal performance.

REFERENCES

1.-Sorab, J., Korcek, S., and Bovington, C., "Friction Reduction in Lubricated component Trough Engine Oil Formulation" SAE Technical paper 982640, 1998, doi: 10.4271/982640.
2.-B.A. Khorramian, G.R. Iyer, S. Kodali, P. Natarajan and R. Tupil, "Review of antiwear additives for crankcase oils" Wear, 169 (1993) 87-95.
3.-C. Muratore and A.A. Voevodin, "Chamaleon Coatings: Adaptive Surfaces to Reduce Friction and Wear in Extreme Environments. Annu. Rev. Mater. Res. 2009. 39:297-324.
4.-R. Hauert and J. Patscheider, "From Alloying to Nanocomposites - Improved Performance of Hard Coatings". Advance Engineering Materials 2000, 2, N_o 5.

5.-Kenneth Holmberg, Allan Matthews, Helena Ronkainen, "Coatings tribology – contact mechanism and surface design" Tribology International Volume 31, Issues 1-3, January 1998, pages 107-120.
6.-A. Erdemir, C. Bindal, G.R. Fenske, C. Zuiker, P. Wilbur, " Characterization of transfer layers forming on surfaces sliding against diamond-like carbon" Surface and Coatings Technology 86-87 (1996) 692-697.
7.–Alfred Grill, "Diamond-like carbon: state of the art". Diamond and Related Materials 8 (1999) 428-434.

RARE EARTH OXIDES PYROCHLORE COMPOUNDS BY SOFT CHEMISTRY

A. Joulia[a,b,c], E. Renard[c], D. Tchou-Kien[c] , M. Vardelle[b] and S. Rossignol[a]

[a] Groupe d'Etude des Matériaux Hétérogènes (GEMH-ENSCI) Ecole Nationale Supérieure de Céramique Industrielle, 12 rue Atlantis, 87068 Limoges Cedex, France.
[b] SPCTS, UMR CNRS no. 6638, Université de Limoges, 12 rue Atlantis, 87068 Limoges Cedex, France.
[c] CNES (French Space Agency), 52 rue Jacques Hillairet, 75612 Paris Cedex, France.
■ Corresponding author - sylvie.rossignol@unilim.fr - tel: 33 5 87 50 25 64

ABSTRACT

Rare-earth zirconates ($Re_2Zr_2O_7$, Re= rare element) have attracted particular interest in thermal barrier coating (TBC) applications due to suitable thermal properties such as low conductivity and efficient phase stability at elevated temperatures. Among them, lanthanum zirconate ($La_2Zr_2O_7$ - LZ) with a pyrochlore structure has been proposed as potential candidate. Recently, rare-earth cerates, especially lanthanum cerate ($La_2Ce_2O_{7-\delta}$ - LC) has been also developed and considered as new TBC ceramic material. This study is interested in synthesizing these compounds by soft chemistry processes: alkoxide and citrate synthesis route. X-ray diffraction (XRD) was used to analyze the powder after calcinations under air. Pyrochlore and fluorite structures were identified at relatively low temperature (1000°C) for LZ and LC respectively. Chemical reactivity tests under reducing atmosphere were performed at 1400°C and investigated by XRD analysis. The lanthanum zirconate is the most stable and interesting material under $Ar_{(g)}/2\%H_{2(g)}$ atmosphere.

1. INTRODUCTION

Advanced turbine engines are being developed to operate at higher temperatures and thereby to increase thermal efficiency in aeronautic, marine and nuclear applications, and particularly for aircraft propulsion [1]. To protect the hot-section metallic components from combustion environments and to preserve thermo-mechanical performances of superalloys at elevated temperatures, thermal barrier coatings are developed as essential [1]. Conventional TBC systems consists of a thermal insulation layer deposed by atmospheric plasma spraying or EB-PVD process on a metallic bond coat (generally MCrAlY, M=Ni/Co) or on a Ni-based superalloy substrate [1, 2]. The most used ceramic top coat is yttria stabilized zirconia (YSZ) in industrial TBC systems, which is able to operate at temperatures below 1200°C. At higher temperatures, phase transformations and accelerated sintering of YSZ lead to early spallation failure of TBC [3]. Thus, the development of new advanced ceramic materials is now manifesting. Among the potential candidates for TBCs such as: doped zirconia [4], complex perovskites [5], aluminates [6, 7], the pyrochlore structured rare-earth zirconates ($Re_2Zr_2O_7$, Re = La, Nd, Sm and Gd) represent alternative and promising ceramic materials [8, 9]. Lanthanum zirconate ($La_2Zr_2O_7$ - LZ) has been proposed as favorable TBCs material because it has some attractive properties such as high melting point (2300°C), low thermal conductivity (1.56 $W.m^{-1}.K^{-1}$ at 1000°C), good phase stability and high sintering resistance [10, 11]. However, LZ has a low thermal expansion coefficient (TEC) compared to yttria stabilized zirconia [11]. Generally, materials containing CeO_2 have usually higher TECs.

Thus, the latter may be improved by the substitution of ZrO_2 by CeO_2. That is why, the rare earth cerates are also interesting candidates for TBC applications and particularly lanthanum cerate ($La_2Ce_2O_{7-\delta}$). The latter has lower thermal conductivity (0.6 $W.m^{-1}.K^{-1}$ at 1000°C) and no phase transformation after long-term annealing at 1400°C [11]. $La_2Ce_2O_{7-\delta}$ is a solid solution of La_2O_3 in CeO_2 with a fluorite structure [12]. Another aspect of thermal barrier coatings must be that they operate and be resistant in oxidizing and reducing combustion environments. To simulate a reducing atmosphere, a gas mixture of $Ar_{(g)}/H_{2(g)}$ is used in our laboratory.

The purpose of this work is to synthesize $La_2Zr_2O_7$ and $La_2Ce_2O_{7-\delta}$ compounds by soft chemistry processes and to investigate their chemical reactivity at 1400°C under air and under particularly reducing atmosphere using XRD analyses.

2. EXPERIMENTAL PROCEDURE

2.1. *Powder synthesis*

$La_2Zr_2O_7$ and $La_2Ce_2O_{7-\delta}$ compounds were synthesized by two soft chemistry processes: alkoxide [13] and citrate synthesis route [14, 15], illustrated in Fig. 1.
The first route was used to synthesize the lanthanum zirconate. The precursors selected were the lanthanum nitrate ($La(NO_3)_3.6H_2O$, purity: 99.9%, Alfa Aesar) and the zirconium n-propoxide (($Zr(OC_3H_7)_4$, purity: 70%, Sigma Aldrich) dissolved in distilled water and n-propanol, respectively. The aqueous solution was added to the zirconium alkoxide solution, which led immediately to the hydrolysis of the mixture and the formation of a pseudo-gel. The latter was dried at 70°C for 48h in a sand bath, grinded in an agate mortar and calcined at 1000°C for 4h under air.

Lanthanum nitrate ($La(NO_3)_3.6H_2O$), cerium nitrate ($Ce(NO_3)_3.6H_2O$), purity: 99.5%, Alfa Aesar), zirconyl nitrate ($ZrO(NO_3)_2.xH_2O$), purity: 99%, Sigma Aldrich) were chosen as reactants for the citrate synthesis method. Stoichiometric amounts of La and Ce or Zr nitrates were dissolved in distilled water. After dissolution of all salts, citric acid ($C_6H_8O_7.H_2O$, purity: 99.5%, Alfa Aesar) was added. The pH value of the solution was maintained to a basic pH. The solvent was evaporated at 80°C until the gel formation, which is dried at 200-250°C during 2h. The amorphous precursor was calcined at 1000°C for 4h under air.

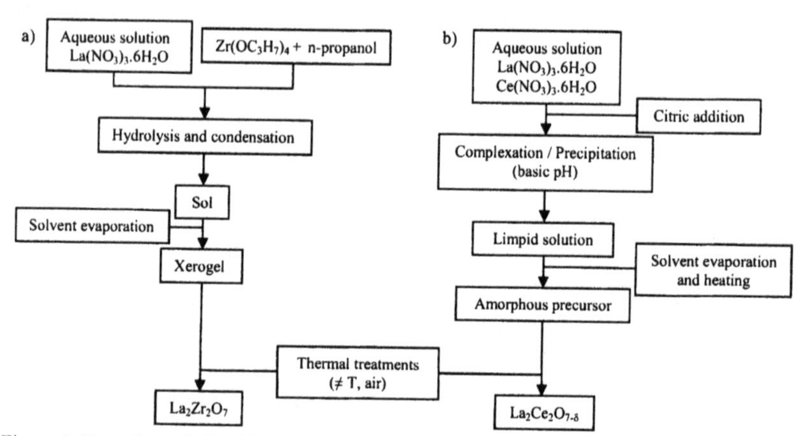

Figure 1: Procedure of alkoxide (a) and citrate (b) synthesis route

2.2. Characterization techniques

XRD analyses of powders were carried out on a Siemens D5000 powder diffractometer using $Cu_{K\alpha}$ radiation ($\lambda_{K\alpha 1}$=1.5406 Å). XRD patterns were acquired using the following parameters: dwell time: 1s; step: 0.045 (2θ). Crystalline phases were identified using Diffrac^{Plus} Eva software by comparison with the database of International Centre of Diffraction DATA JCPDS-ICDD®. Scherrer analysis was performed in order to assess the average crystalline domain size. The latter is calculated by the Scherrer equation with FWHM and peak position (2θ) of several peaks. The average value is given as the mean crystallite size.

Thermogravimetric (TG) and differential thermal (DTA) experiments were performed on a SDT Q600 apparatus from TA instrument in a Pt crucible between 25°C and 1200°C. The samples were heated at 10°C.min^{-1} in a dry air flow (100 mL.min^{-1}).

Specific surface areas were determined by N_2 adsoption at -196°C (one point Brunauer-Emmet-Teller (BET) method) using a Mircromeritics Tristar 2. The powder samples are degased at 200°C for 12h. The densities were measured by a helium pycnometer (AccucPyc 1330 - Micromeritics).

The chemical reactivity of selected materials was carried out in a furnace under reducing atmosphere at 1400°C for 4h in a gas mixture of $Ar_{(g)}/2\%H_{2(g)}$ (flow rate: 500 mL.min^{-1}). A heating linear ramp rate of 5°C.min^{-1} until 1400°C was used. The cooling rate was fixed to 5°C.min^{-1}. Afterwards, the samples were analyzed by XRD.

The sample nomenclature is reported in Table 1 in function of synthesis mode.

Table 1: Features of $La_2Zr_2O_7$ and $La_2Ce_2O_{7-\delta}$ after a heating treatment at 1000°C (* 1400°C under air)

Sample	Chemistry process	Nomenclature	Density (g.cm^{-3})	Specific area (m^2.g^{-1})	BET size (nm)	XRD size (nm)
$La_2Zr_2O_7$	Alkoxide route	LZ-A	5.7	5	105	12 ± 2 *120 ± 2
	Citrate route	LZ-C	6.2	10	48	26 ± 2 *131 ± 2
$La_2Ce_2O_{7-\delta}$	Citrate route	LC-C	6.3	10	47	36 ± 2 *136 ± 2

3. RESULTS AND DISCUSSION

3.1. Specifications of different compounds

3.1.1. $La_2Zr_2O_7$ compound

Thermal analysis was only used to characterize the xerogel prepared by alkoxide route. The TGA-DTA results are plotted in Fig. 2. The thermogravimetric curve shows a total weight loss of 45% starting at 50°C and carried on until 500°C associated with endothermic peaks. The endothermic peaks at 100°C and 430°C, are attributed to physical water and to the decomposition of residual precursors due to removal organics, respectively. At 870°C, the exothermic peak, without weight loss, can be attributed to a crystallization phenomenon.

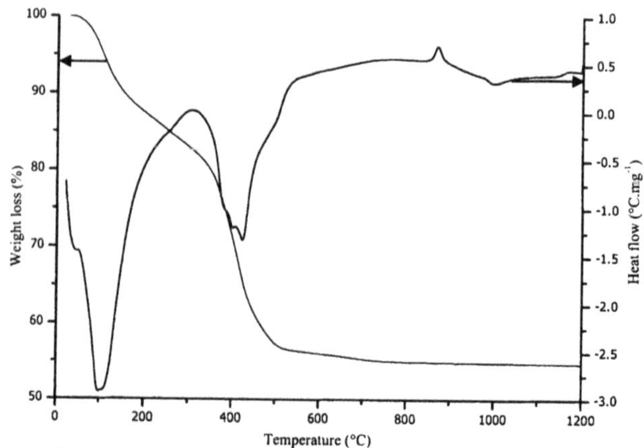

Figure 2: Thermal analysis of xerogel (La-O-Zr) synthesized by alkoxide route

In order to follow the crystallization of $La_2Zr_2O_7$ compound in function of temperature, successive calcinations of powder samples were carried out. The XRD patterns of xerogel calcined under air at different temperatures (800°C, 900°C, 1000°C and 1400°C) are presented in Fig. 3. At 800°C, the diffraction lines corresponding to (220), (400) and (440) suggest the formation of the lanthanum zirconate with a fluorite type structure (Fm3m) in agreement with literature data [16]. Increasing the temperature to 900°C, the diffraction peaks become narrower and sharper due to crystallization phenomenon. At 1000°C, the presence of the super-lattice peaks such as (111), (311) and (511) is generally attributed to pyrochlore structure in a cubic space group (Fd3m). This transformation is in agreement with the small exothermic peak detected in the DTA curve around 1000°C without weight loss. At 1400°C, no secondary phase is detected by XRD analysis, which evidences that the pyrochlore phase is relatively stable (JCPDS file No. 01-070-5602). The same results (Fig. 3e and f) are evident by using the citrate route since the pyrochlore phase remains stable until 1400°C.

From the specific area and the density of powders and assuming the crystallites are spherical, the crystallite size can be calculated. The values are reported in Table 1. After calcinations at 1000°C, the BET sizes are higher to XRD sizes revealing that the powders are agglomerated. The thermal treatment at 1400°C causes powder sintering and grain coalescence according to XRD sizes: 121 nm and 131 nm for LZ-A and LZ-C, respectively.

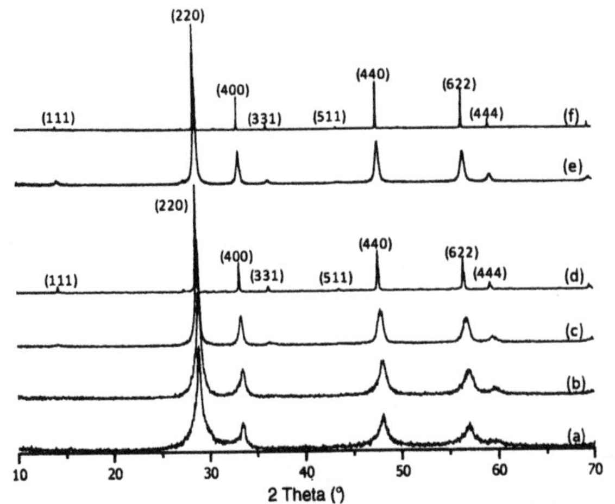

Figure 3: XRD patterns for LZ-A at a temperature of (a) 800°C, (b) 900°C, (c) 1000°C, (d) 1400°C and for LZ-C at (e) 1000°C and (f) 1400°C (JCPDS file No. 01-070-5602)

3.1.2. $La_2Ce_2O_{7-\delta}$ compound

Fig 4 shows the XRD patterns of LC amorphous precursor with different thermal treatments. At 1000°C, $La_2Ce_2O_{7-\delta}$ compound displays a fluorite structure belonging to cubic space group (Fm3m) whatever the calcination temperature (JCPDS file No. 04-012-6396). The use of δ symbol was preferred since these oxides are non-stoichiometric due to the oxygen vacancy in relation with cerium valence (III/IV) The comparison of the size determined by XRD or BET reveals a weak agglomerated state. At 1400°C, the crystallite size is 136 nm, which also confirms the grain sintering (36 nm at 1000°C).

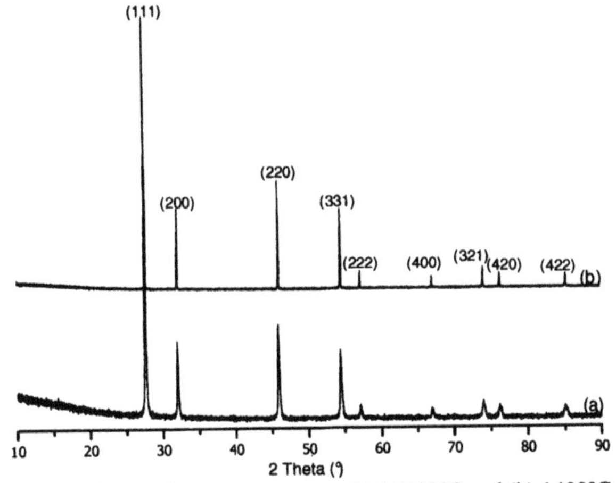

Figure 4: XRD patterns for LZ-C at a temperature of (a) 1000°C and (b) 1400°C (JCPDS file No. 04-012-6396)

3.2. *Reactivity under reducing atmosphere at 1400°C*

After, the chemical reactivity of $La_2Zr_2O_7$ and $La_2Ce_2O_{7-\delta}$ compounds faced to hydrogen atmosphere and under low partial pressure of oxygen at high temperature, XRD experiments were performed.

Fig. 5A presents the XRD patterns of LZ-A and LZ-C sample powders after a thermal treatment under $Ar_{(g)}/2\%H_{2(g)}$. Whatever the synthesis process used, the pyrochlore phase is always present. An enlargement of XRD patterns (Fig3 5B) shows some peaks can be attributed to lanthanum hydroxide and to zirconia. These phases could be explained by the water vapor formation in spite of the $Ar_{(g)}/2\%H_{2(g)}$ flow in the furnace. The lanthanum zirconate would be divided into minority phases such as zirconia (ZrO_2) and lanthanum oxide (La_2O_3) according to the equations (1 - 2):

$$La_2Zr_2O_7 \xrightarrow{H_{2(g)}} (1\text{-}x)La_2Zr_2O_7 + xLa_2O_3 + 2xZrO_2 \qquad (1)$$

$$La_2O_3 \xrightarrow{H_2O_{(g)}} 2La(OH)_3 \qquad (2)$$

The lines of diffraction (noted by an asterix) of very weak intensity could be attributed to the presence of a ($La_xSi_yO_z$) compound. It could be formed during the thermal treatment between our compound and the crucible based on silica and alumina.

The XRD pattern (Fig. 6) highlights the presence of several phases such La_2O_3 and Ce_2O_3 suggesting the total decomposition of $La_2Ce_2O_{7-\delta}$ compound under hydrogen. In fact, it is well known that under hydrogen, the cerium compounds are not stable since they have the ability to transform into (Ce_2O_3) due to the reduction of +IV cerium to +III cerium [17]. In fact, in presence of reducing atmosphere, the $La_2Ce_2O_{7-\delta}$ compound is decomposed to cerium oxide (Ce_2O_3) and lanthanum oxide (La_2O_3) (equation 3):

$$La_2Ce_2O_{7-\delta} \xrightarrow{H_{2(g)}} La_2O_3 + Ce_2O_3 + (1\text{-}\delta)/2\ O_{2(g)} \qquad (3)$$

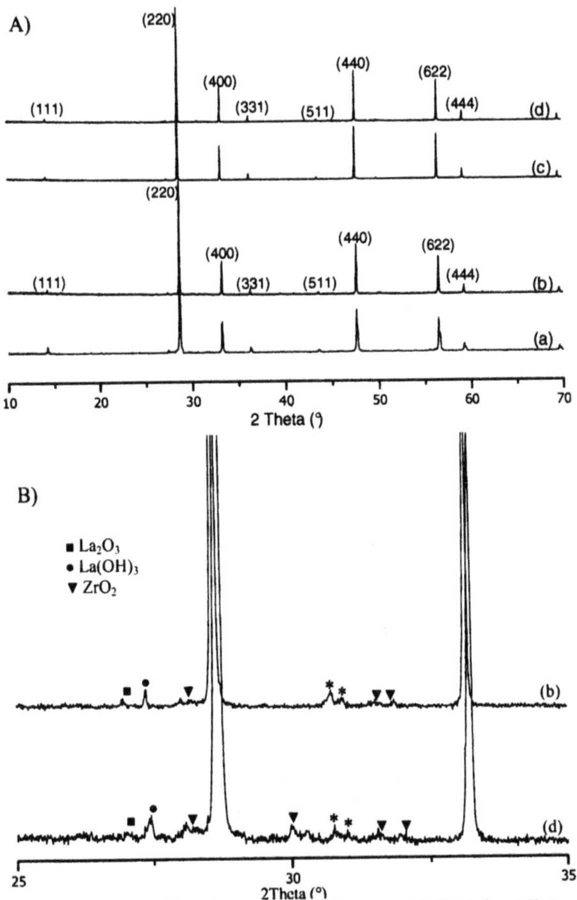

Figure 5: A: XRD patterns as function of atmosphere at 1400°C for LZ-A: (a) air, (b) $Ar_{(g)}/2\%H_{2(g)}$ and for LZ-C: (c) air, (d) $Ar_{(g)}/2\%H_{2(g)}$ and B: XRD patterns enlargement for LZ-A and LZ-C at 1400°C under $Ar_{(g)}/2\%H_{2(g)}$ (La_2O_3: JCPDS file No. 03-065-3185, $La(OH)_3$: JCPDS file No. 01-077-3931, ZrO_2: JCPDS file No. 04-011-8817 and 04-006-9855)

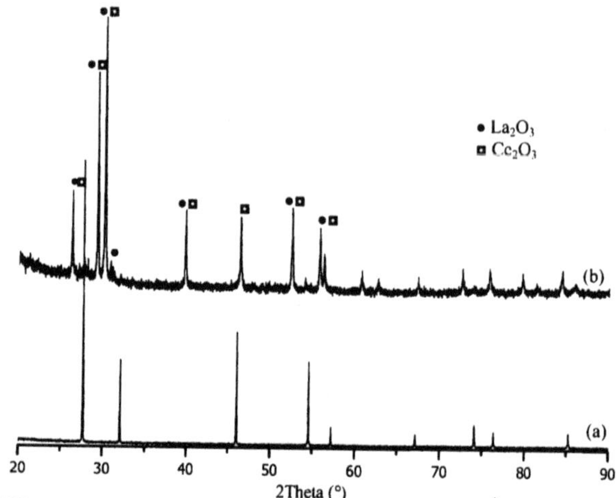

Figure 6: XRD patterns as function of atmosphere at 1400°C for LC-C: (a) air, (b) $Ar_{(g)}/2\%H_{2(g)}$ (La_2O_3: JCPDS file No. 04-015-5007 and 04-002-3877, $La(OH)_3$: JCPDS file No. 01-077-3931, Ce_2O_3: JCPDS Card No. 00-023-1048)

4. CONCLUSION

$La_2Zr_2O_7$ and $La_2Ce_2O_{7-\delta}$ compounds were successfully synthesized by soft chemistry processes at relatively low temperature (1000°C). Pyrochlore and fluorite structures remain stable and no phase decomposition occurs after calcination at 1400°C under air.

The chemical reactivity tests, under reducing atmosphere at 1400°C, reveals that the $La_2Ce_2O_{7-\delta}$ compound is unstable due to valence change of cerium, which results in new phase formation. Whereas, the pyrochlore structure of lanthanum zirconate remaining stable under $Ar_{(g)}/2\%H_{2(g)}$, appears promising for TBC applications in extreme environments. This stability should be also confirmed under water vapor at high temperature.

ACKNOWLEDGEMENTS

The authors would like to thank the French DGA (Ministry of Defense) and the CNES (national R&T program) for their financial support.

REFERENCES

[1] R.A. Miller, Thermal barrier coatings for aircraft engines : history and directions, Journal of Thermal Spray Technology, 6, 35-42 (1997)
[2] N.P. Padture, M. Gell and E.H. Jordan, Thermal barrier coatings for gas-turbine engine applications, *Science*, 296, 280-284 (2002)
[3] J.R. Brandon and R. Taylor, Phase stability of zirconia-based thermal barrier coatings Part I. Zirconia-yttria alloys , *Surface and Coatings Technology*, 46, 75-90 (1991)
[4] D. Zhu and R.A. Miller, Development of advanced low conductivity thermal barrier coatings, *International Journal of Apllied Ceramic Technology*, 1, 86-94 (2004)
[5] R. Vassen, X. Cao, F. Tietz, D. Basun and D. Stöver, Zirconates as New Materials for Thermal Barrier Coatings, *Journal of the American Ceramic Society*, 83, 2023-2028 (2000)
[6] C.J. Friedrich, R. Gadow and T. Schirmer, Lanthanum hexaaluminate - A new material for atmospheric plasma spraying of advanced thermal barrier coatings, *Journal of Thermal Spray Technology*, 10, 592-598 (2001)

[8] J. Wu, X. Wei, N.P Padture, P.G. Klemens, M. Gell, E. Garcia, P. Miranzo and M.I. Osendi, Low-themal-conductivity rare-earth zirconates for potential thermal-barrier-coating applications, *Journal of the American Ceramic Society*, 85, 3031-3035 (2002)
[9] R. Vassen, M.O. Jarligo, T. Steinke, D.E Mack and D. Stöver, *Overview on advanced thermal barrier coatings, Surface & Coatings Technology*, 205, 938-942 (2010)
[10] X. Q. Cao, R. Vassen,, W. Jungen, S. Schwartz, F. Tietz and D. Stover, Thermal Stability of Lanthanum Zirconate Plasma-Sprayed Coating, *Journal of the American Ceramic Society*, 84, 2086-2090 (2001)
X. Q. Cao, R. Vassen, W. Fischer, T. Tietz; W. Jungen and D. Stöver, Lanthanum-cerium oxide as a thermal barrier-coating material for high-temperature applications, Advanced Materials,1438-1442 (2003)
[13] S. Rossignol, Y. Madier and D. Duprez, *Preparation of zirconia-ceria materials by soft chemistry*, Catalysis Today, 50, 261-270 (1999)
[14] Q.Xu, D. Huang, W. Chen, J Lee, H Wang and R. Yuan, Citrate method synthesis, characterization and mixed electronic-ionic conduction properties of $La_{0.6}Sr_{0.4}Fe_{0.2}O_3$ perovskite type complex oxides, *Scripta Materiala*, 50, 165-170 (2004)
[15] A. Julian, E. Juste, P.M. Geffroy, V. Coudert, S. Degot, P. Del Gallo, N. Richet and T. Chartier, H. Chen and Y. Gao, Coprecipitation synthesis and thermal conductivity of $La_2Zr_2O_7$, *Journal of Alloy Compounds*, 480, 843-848 (2009)
S. Rossignol, D. Mesnard, F. Gerard, C. Kappenstein and D. Duprez, Structural changes of Ce-Pr-O oxides in hydrogen: a study in situ X-ray diffraction and Raman, *Journal of Materials. Chemistry*, 13, 3017-3020 (2003)

DEGRADATION OF (Ni,Pt)Al COATINGS BY MIXTURE OF SODIUM AND POTASSIUM SULFATE AT 950°C

Le Zhou, Sriparna Mukherjee, Yongho Sohn
Advanced Materials Processing and Analysis Center/ Department of Mechanical, Materials and Aerospace Engineering, University of Central Florida, Orlando, FL, United States

ABSTRACT

Environmental stability of β-(Ni,Pt)Al coatings due to combustion by-product of fuel impurities has been examined owing to recent interests in bio-derived (e.g., algae-derived) fuels. Pure sodium sulfates (Na_2SO_4), potassium sulfates (K_2SO_4) and three of their mixtures with different weight ratio were prepared by mechanically grinding, and their high temperature interactions with diffusional Pt-modified β-NiAl coatings were investigated in a laboratory furnace at 950°C. The corroded samples were characterized by X-ray diffraction and scanning electron microscopy equipped with X-ray energy dispersive spectroscopy. The results showed severe damages occurred in the β-(Ni,Pt)Al coatings when the salts were in molten state at 950°C via fluxing mechanism, and accelerated oxidation also occurred when the pure solid state K_2SO_4 was in contact with β-(Ni,Pt)Al.

INTRODUCTION

Increasing demands for higher operating temperature of gas turbines continuously drive the development of thermal barrier coatings (TBCs), which serve as the high temperature protector for the underlying hot components. Current state-of-art TBCs provide a temperature gradient of about 150°C, which allow the superalloys to operate above their melting temperature [1]. However, the degradation issue of the coatings under harsh environment greatly inhibits the long-term stability and lifetime of the TBC system. Conventionally, the degradation is divided into three kinds according to the temperature that the coatings are exposed to: oxidation (>1000°C), Type I hot corrosion (800-1000°C) and Type II hot corrosion (600-800°C) [2-4].

Pt-modified β-NiAl has been widely applied on gas turbines as the bond coats due to its excellent ability of forming protective alumina scale under high temperature. From numerous research that have been performed up to now, the addition of Pt into β-NiAl is confirmed to be beneficial to the oxidation behavior primarily because of the slowed diffusion of detrimental element, improved alumina scale adhesion and retarded void growth at the interface [5, 6]. It was also pointed out that the Type II hot corrosion behavior of β-(Ni,Pt)Al coatings is much better than that of Co based metallic overlay coatings. The latter ones tend to form the Na_2SO_4-$CoSO_4$ eutectic (T_m=565°C) under Type II hot corrosion condition. However, β-(Ni,Pt)Al can be severely damaged under Type I hot corrosion condition. Since future advanced aero or industrial engines operate at both hot corrosion and oxidation environment, further improvement of β-(Ni,Pt)Al becomes urgent. In a recent study by Task and Gleeson, 5% of Co addition into Ni-36Al-5Pt improved the Type I hot corrosion resistance, which was explained by the prevention of internal Al oxidation [7]. Deodeshmukh and Gleeson also produced (Pt,Hf)-modified γ'-Ni_3Al + γ-Ni alloys which showed improved Type I hot corrosion behavior, due to the slow growing and more adherent alumina scale [3, 8].

Previous studies on the hot corrosion tests usually employed Na_2SO_4 as the base corrosive salts [7, 9]. This is because Na is easily ingested when the turbines operate in the marine atmosphere and is also contained in conventional fuel [10]. Literature reviews show that some studies added K_2SO_4, NaCl or both of them into Na_2SO_4 up to a weight percent of 25% to form a mixed corrosive salts with lower melting point, thus expanding the corrosion range [11-13]. However, the effect of K_2SO_4 on the hot corrosion behavior of the β-(Ni,Pt)Al has not been clarified since K_2SO_4 behaviors were believed to be similar to Na_2SO_4 as far as Type I hot corrosion was concerned. Recent interests in the second generation of biofuel extracted from algae require the examination of compatibility between the biofuel and the high temperature coatings. The majority impurities of the algal biomass include Na, K, P, Si, S and Cl [14, 15]. Therefore, it is also necessary to identify K_2SO_4 as a major corrosive salt and investigate its effect on hot corrosion.

In this study, cyclic hot corrosion behavior of commercial β-(Ni,Pt)Al coatings at 950°C was investigated. The corrosive salts applied in this study involve Na_2SO_4, K_2SO_4 and three kinds of their mixtures with different weight ratio. The effect of these corrosive salts on the hot corrosion behavior was compared and discussed with respect to the mechanism of corrosion.

EXPERIMENTAL

The specimens used in this experiment were commercially available Pt-modified NiAl coatings applied on the CMSX-4 superalloy substrate (Howmet Research Corporation). The deposition process of Pt-modified NiAl involves the electroplating of a Pt layer onto the substrate and subsequent high-temperature low-activity aluminizing via chemical vapor deposition (CVD), where Ni grows outward from the substrate to form a single β phase coating. The nominal composition of the CMSX-4 substrate (in weight percentage) is Ni-11Co-5.7Cr-5.6Al-5.6Ta-5.2W-3Re-1Hf-0.8Ti-0.6Mo.

For the laboratory hot corrosion tests, Na_2SO_4 (T_m=884°C), K_2SO_4 (T_m=1069°C) and three of their mixtures were chosen as the corrosion deposits. These mixtures were prepared by weighing Na_2SO_4 and K_2SO_4 at ratios of 3:1, 1:1 and 1:3 using electronic balance and mechanically grinding into fine powders. A disc specimen with diameter of 25.4mm was sectioned into 6 equal pieces (i.e., surface area) by a low speed diamond saw. Each of the pieces was uniformly coated with one kind of corrosive salts at a concentration of ~10 mg/cm^2 by using a clean and dry glass rod, and one piece was left uncoated for ambient oxidation experiment. For convenience, these specimens were marked as samples 1 through 5 in the sequence of increasing amount of K_2SO_4 and sample 6 for the one without salt. Hot corrosion tests were conducted using a high temperature laboratory furnace at 950°C. The procedure consisted of heating the specimens to 950°C in 15 min, isothermally dwelled for 10 hours followed by furnace cooling to room temperature. Specimens were taken out from the furnace after each cycle and weighed by electronic balance without any attempt to retain the spalled oxides from the surface. The specimens were recoated with the same amount of salts after each cycle. Totally 4 cycles were carried out before severe coating spallation occurred by visual examination.

Characterization of the surfaces and cross-sections of the as-coated specimen and the corroded specimens was carried out by using X-ray diffraction (Rigaku DMax-B) and scanning electron microscope (Zeiss 55 Ultra-55 field emission SEM) equipped with X-ray energy dispersive spectroscopy (XEDS) with Si drift detector. All specimens were sectioned by a low speed saw, mounted in epoxy, ground and polished to the final stage of ~ 0.25μm, and ultrasonically cleaned by acetone and ethanol before the cross-sectional SEM analysis.

RESULTS AND DISCUSSIONS

Microstructural Characteristics of As-coated β-(Ni,Pt)Al

The results from the characterization of as-coated β-(Ni,Pt)Al coatings are presented in Figure 1 (a-d). The XRD analysis showed exclusively β phase of the as-coated coatings as illustrated in Figure 1 (c). The surface morphology has a typical "ridge" microstructure with the aluminide grains (~30-60 μm) extending from the coating/substrate interface to the coating surface. From the cross-sectional micrograph, both the outer part of the β-(Ni,Pt)Al and the interdiffusion zone have a thickness of around 25 μm. Such microstructure results from the outward diffusion of Ni from the substrate to the coating during aluminizing process, whereas the refractory elements such as W, Mo, Ta and Re precipitate out from the Ni solution to form the interdiffusion zone. The average composition measured by XEDS for the outer portion of the β-(Ni,Pt)Al was approximately, in weight percent, 42.6Al-36.8Ni-6.5Pt-4.8Co-2.8Cr-0.7Ti.

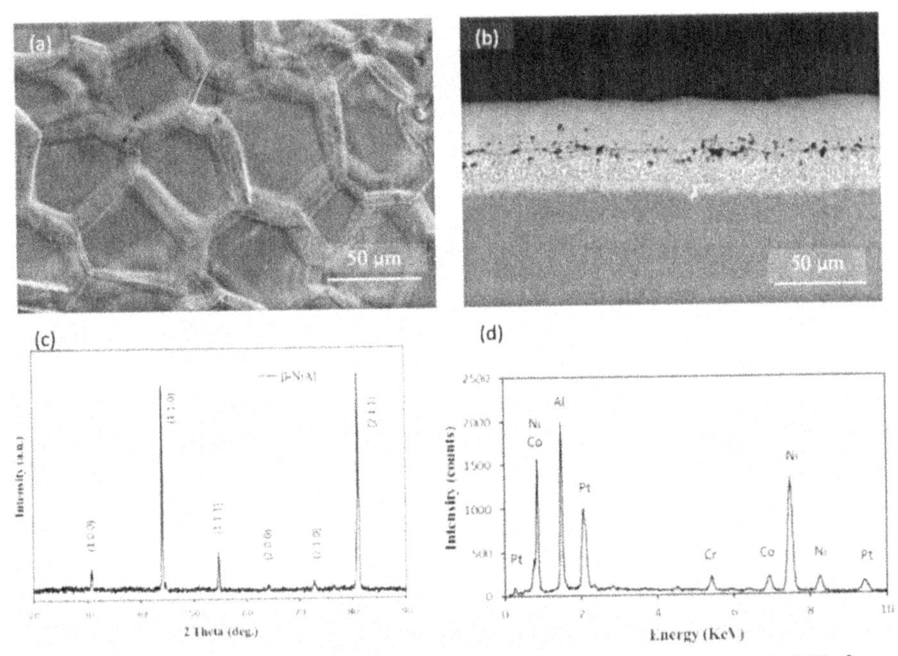

Figure 1: Microstructural characteristics of the as-coated Pt-modified NiAl coatings: (a) SEM of surface morphology (b) backscatter electron cross-sectional micrograph (c) XRD pattern (d) XEDS from outer portion of β-(Ni,Pt)Al coatings.

Mass Gain Analysis

Weight gains of all the specimens during the tests were documented after each cycle, and the comparison of the weight gains for the 6 samples after 40h of tests is presented in Figure 2. For samples 1-4, the weight gain due to salt deposition was subtracted in the calculation, while for sample 5 this was not considered because K_2SO_4 remained solid state and automatically fell off from the sample during handling. From Figure 2, sample 6 that was exposed to only pure air had a very small mass gain of about 0.25 mg/cm^2. This is because the temperature for oxidation in this test was only 950°C and the exposure time of 40h is relatively short. It also, to some degree, confirms that the growth rate of oxides on this β-(Ni,Pt)Al coating is slow and thus is a proof of good resistance to oxidation. For samples 1-4, the scope of mass gains ranged from 45-55 mg/cm^2, indicating that a heavy corrosion occurred. In some previous research, Deodeshmukh and Gleeson observed a mass gain of about 24 mg/cm^2 and 19 mg/cm^2 for $\gamma+\gamma'$ Ni-22Al-5Pt-0.4Hf and Ni-22Al-10Pt-0.4Hf, respectively, after hot corrosion at 900°C for 100h by applying 1.5 mg/cm^2 Na_2SO_4 every 20 hours [8]. Task and Gleeson obtained a mass gain of 17 mg/cm^2 for Ni-36Al-5Pt after 180h Type I hot corrosion with deposition of 2-3 mg/cm^2 Na_2SO_4 every 20h cycle [7]. Considering the different parameter used in our experiment, that is a larger deposit concentration and a shorter interval of salt recoating, the higher weight gain is reasonable. Also, the relative large amount of salts used in this study dripped off from the surface and corroded the substrate edge (small area compared to the coating surface), which might also contribute to the mass gain of samples 1 to 4. The melting temperature of Na_2SO_4 and K_2SO_4 is 884°C and 1069°C, and the salts deposited in samples 2, 3 and 4 begin to melt at 852°C, 862°C and 912°C, respectively [16]. Sample 5 exhibited a relatively small mass gain of about 2.0 mg/cm^2 after 40 hours corrosion. This implies that corrosion did occur although the pure K_2SO_4 remained solid (e.g., compared to pure oxidation in sample 6). However, the damage in sample 5 was much less severe compared to the damage in samples 1 through 4.

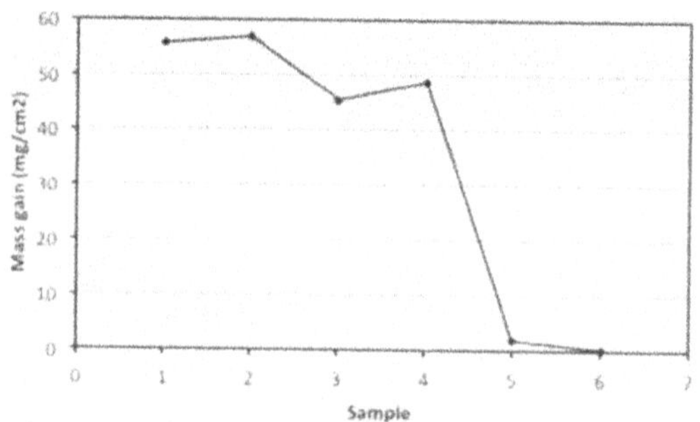

Figure 2: Weight gains per unit area after 40 hours of hot corrosion (samples 1 through 5) and oxidation (sample 6) tests at 950°C.

Phase Constituents of Hot Corrosion Products

Figure 3 represents the XRD patterns from samples 1 through 6 after 40 hours of tests. XRD pattern from the as-coated β-(Ni,Pt)Al coatings is presented for comparison. For sample 6, Al_2O_3 and β-(Ni,Pt)Al phases were detected, indicating that the α-Al_2O_3 scale developed on the surface. For samples 1 through 5, the primary phases consisted of Al_2O_3 and γ'-Ni_3Al with some traces of the remaining β phase. This indicated that the Al was rapidly consumed during the hot corrosion tests, which resulted in the formation of Al_2O_3 scale as well as the depletion of Al in β phase. There were also some minor peaks that were unidentified and inconsistent, which could be oxides of chromium or titanium inferred from XEDS data of both the surface and cross-section. Also, no sulfides peaks were seen in the XRD pattern whereas sulfides were frequently identified later by XEDS. This might be attributed to the fact that sulfides developed beneath the Al_2O_3 scale.

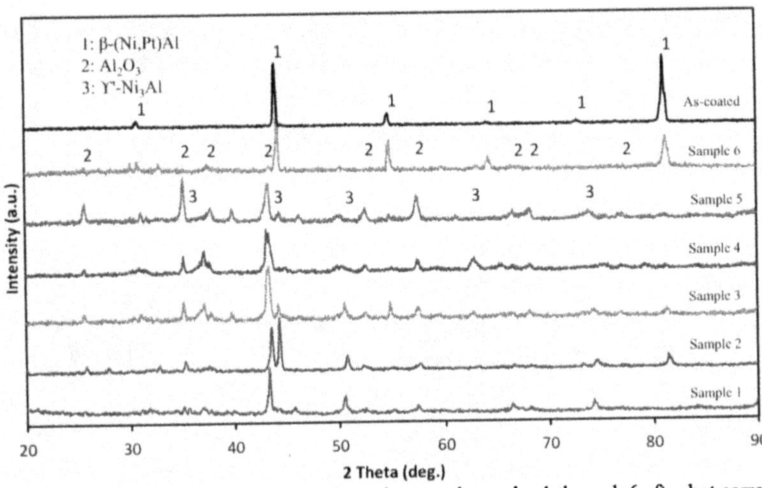

Figure 3: X-ray diffraction patterns of as-coated specimen and samples 1 through 6 after hot corrosion and oxidation tests at 950°C for 40 hours.

<u>Microstructure and Mechanisms of Hot Corrosion</u>

Figure 4 shows the cross-sectional micrographs of samples 1 through 6 after 40 hours of high temperature exposure at 950°C. The overall microstructure of the 6 cross-sectional micrographs can be divided into three groups, with samples 1 through 4 having similar damaged microstructure, sample 5, and sample 6 exhibiting slightly different damaged microstructure. Figure 4 (f) shows that a thin layer of thermally grown oxide (TGO) with a thickness of about 3 μm formed on the surface of sample 6 after 40 hours of oxidation under ambient environment. XEDS analysis on both the surface and cross-section indicated that the TGO mainly consisted of Al_2O_3 with trace amount of Cr- and Ni-oxides. This uniform and dense TGO, and its slow growing nature observed by the mass gain shown in Figure 2 confirms that β-(Ni,Pt)Al behaves well in resisting high temperature oxidation.

Figure 4 (a through d) present cross-sectional micrographs of samples 1 through 4, respectively, coated with salts having melting temperatures below 950°C. All of these samples were severely attacked by the corrosive salts. The external surface of the samples consisted of a porous oxide layer with thickness of 5 to 10 μm. This oxide layer, examined by XEDS, was comprised of mostly Al_2O_3 and small amount of Ti- and Cr-oxides. Sodium and potassium were also detected by XEDS data in the oxide layer but not in the bond coat layer, which suggested the formation of aluminate or chromate. Internal sulfidation was observed in these four samples as shown by the grey phases in Figure 4 (a through d). The phenomena of sulfidation occurred beneath the oxides layer, and sulfides formed throughout the coatings. Sulfidation even penetrated into the interdiffusion zone as can be seen by the continuous sulfides layer just above the interdiffusion zone.

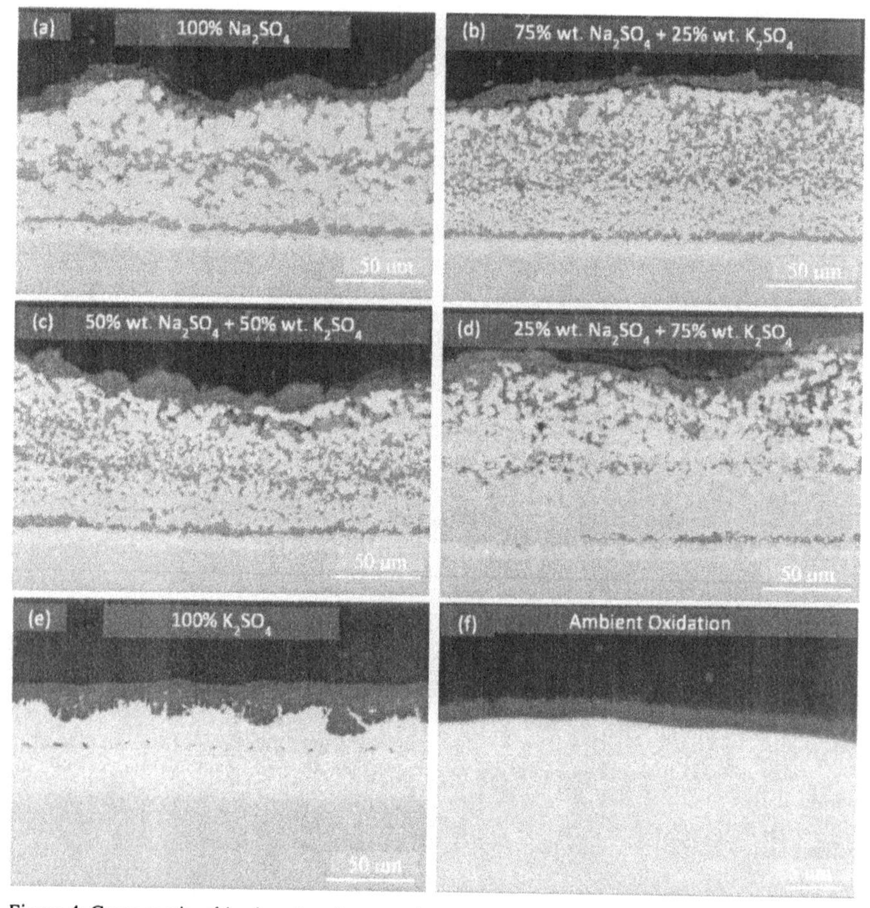

Figure 4: Cross-sectional backscatter electron micrographs (a through f) corresponding to samples 1 through 6 after hot corrosion and oxidation tests at 950°C for 40 hours.

The microstructure of samples 1 through 4 after hot corrosion corresponds to that of traditional Type I hot corrosion and the mechanism would follow the basic fluxing model proposed by Goebel and Pettit [17, 18]. The essence of this model is the dissolution of protective oxides in the salt/metal interface and re-precipitation as porous and non-protective oxides at the salt/gas interface. Oxygen diffuses through the molten salt layer and causes the initial oxidation of aluminum, inducing a concentration gradient of oxygen across the molten salt layer where oxygen activity is lowered at the salt/metal interface. Molten salt becomes unstable in the reduced oxygen environment and decomposes according to the following reaction:

$$SO_4^{2-} = \frac{3}{2} O_2 + O^{2-} + S \tag{1}$$

Both O^{2-} and S activity increase as a result. The sulfur diffuses into the coatings and reacts with Ni, Al and Cr, and results in sulfide formation as seen by Figure 4 (a-d). The O^{2-} reacts with alumina to form aluminate at the salt/oxide interface and diffuses into the molten salt layer:

$$Al_2O_3 + O^{2-} = 2AlO_2^{2-} \quad (2)$$

The aluminate would again precipitate out as alumina particles near the salt/gas interface where the O^{2-} activity is lower. The result is the formation of a porous and non-uniform oxide layer at the surface as shown in Figure 4 (a-d). This oxide is not as effective as that shown in Figure 4 (f) in preventing the penetration of oxygen, and thus an accelerated oxidation process occurs.

Figure 5 (a) presents the composition of the sulfide phases and indicates that Cr, Al and Ni sulfides are the primary sulfides. The formation of sulfides greatly affects the performance of the β-(Ni,Pt)Al coatings. Firstly, sulfur enters into the coatings and reacts with Al and Cr that are important oxide formers, causes accelerated depletion of these elements, and jeopardizes the ability of forming protective oxides. This is presented in Figure 5 (b) where the amount of Al and Cr remaining in the coating decreased substantially compared to that of the as-coated presented in Figure 1 (d). The formation of protective Al_2O_3 on the surface of β-(Ni,Pt)Al is essential in resisting hot corrosion. However, the combination of Al and S reduces the amount of Al in forming oxides, as well as the ability of repairing the spalled oxides. In addition, sulfide formation damages the mechanical integrity of the β-(Ni,Pt)Al. It is recalled that the original thickness of the coating (interdiffusion zone excluded) was 25 μm, while after hot corrosion, the thickness of coating along with corrosion products on top increased to about 80-100 μm. In Figure 6, a secondary electron micrograph for the surface morphology after hot corrosion, together with the cross-sectional micrographs from Figure 4 (a-d), illustrated that there were large surface undulations present on the surface, which can be attributed to sulfide formation.

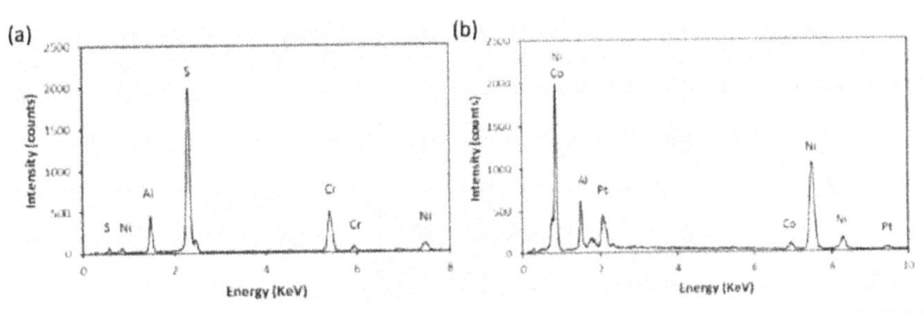

Figure 5: X-ray energy dispersive spectroscopy of: (a) grey phases of sulfidation of Al, Cr and Ni; (b) composition of coatings after corrosion implying accelerated Al depletion.

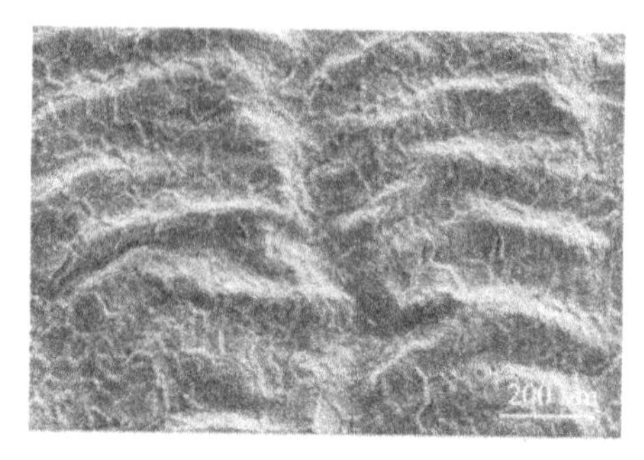

Figure 6: Typical secondary electron micrograph from the surface from samples 1 through 4 after 40 hours of hot-corrosion tests.

Figure 4 (e) presents the cross-sectional microstructure of β-(Ni,Pt)Al coatings after exposure to pure K_2SO_4 at 950°C for 40 hours. The dark contrast phase formed on the upper part of the coating contained mostly Al and O with a very small amount of Cr. However the oxide/metal interface appeared non-planar and the depth of attack varied between 10 to 20 μm. Beneath the oxides layer, large area of lighter contrast phase contained less amount of Al compared with that of the β phase, implying the formation of γ' phase, and a faster consumption of Al compared to pure oxidation in ambient at 950°C. The oxidation of β-(Ni,Pt)Al coating was accelerated due to the presence of K_2SO_4 but not in the mechanism observed for hot-corrosion degradation seen in samples 1 through 4.

Visual examination of the salt after each 10 hours cycle showed that the salt remained in solid state with slight discoloration. It is possible that there was a formation of chromates during the reaction at this temperature. Figure 7 presents the secondary electron micrograph of the surface of sample 5 exposed to pure K_2SO_4 950°C for 40 hours. The irregular surface with pitting attack everywhere is different from any other samples examined in this study. Although the detailed mechanism is currently unknown, it is clear that the K_2SO_4 did cause accelerated oxidation to β-(Ni,Pt)Al coatings with the development of non-planar oxide/metal interface.

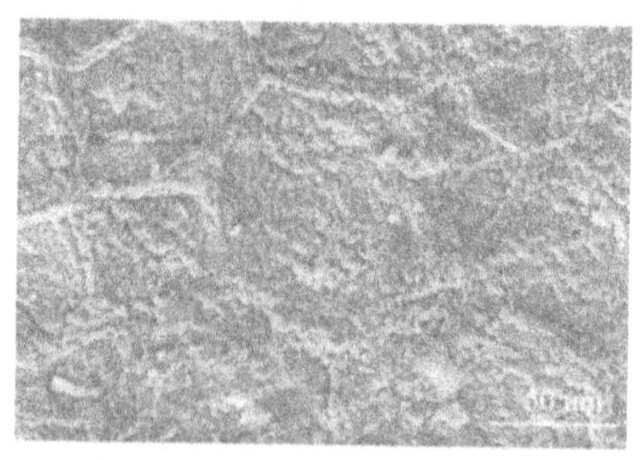

Figure 7: Secondary electron micrograph of the β-(Ni,Pt)Al surface of sample 5 exposed to pure K_2SO_4 at 950°C for 40 hours.

CONCLUSION

In this study, hot corrosion test on diffusional β-(Ni,Pt)Al was performed with different corrosive salts and under ambient environment. A continuous and dense aluminum oxides formed on the surface when only oxidation occurred, indicating good resistant to oxidation. Na_2SO_4 and mixed salts caused severe damage to the β-(Ni,Pt)Al coatings by greatly accelerating the oxidation process. Sulfides formed throughout the β-(Ni,Pt)Al coatings and oxides were dissolved in the molten salts and re-precipitated out as a non-protective oxide. K_2SO_4 also accelerated the oxidation process although it remained solid at 950°C.

ACKNOWLEDGEMENTS

Authors would like to sincerely thank the financial support from the U.S. Air Force Research Laboratory (FA8650-11-C-2127) for the program, "Algal Biofuels for Aviation," administered by New Mexico State University, Las Cruces, NM.

REFERENCES

[1]N. P. Padture, M. Gell, and E. H. Jordan, Thermal barrier coatings for gas-turbine engine applications, *Science*, **296**, 280, 2002

[2]M. Pomeroy, Coatings for gas turbine materials and long term stability issues, *Materials & Design*, **26**, 223-231, 2005

[3]V. Deodeshmukh, N. Mu, B. Li, et al, Hot corrosion and oxidation behavior of a novel Pt+Hf-modified γ' -Ni3Al+ γ -Ni-based coating, *Surface and Coatings Technology*, **201**, 3836-3840, 2006

[4]F. Pettit, Hot Corrosion of Metals and Alloys, *Oxidation of Metals*, **76**, 1-21, 2011

[5]C. Leyens, B. Pint, and I. Wright, Effect of composition on the oxidation and hot corrosion resistance of NiAl doped with precious metals, *Surface and Coatings Technology*, **133**, 15-22, 2000

[6]Y. Zhang, J. Haynes, W. Lee, et al, Effects of Pt incorporation on the isothermal oxidation behavior of chemical vapor deposition aluminide coatings, *Metallurgical and Materials Transactions A*, **32**, 1727-1741, 2001

[7]M. N. Task, B. Gleeson, F. S. Pettit, et al, Compositional effects on the Type I hot corrosion of β-NiAl alloys, *Surface and Coatings Technology*, **206**, 1552-1557, 2011

[8]V. Deodeshmukh and B. Gleeson, Effects of Platinum on the Hot Corrosion Behavior of Hf-Modified γ' -Ni3Al + γ -Ni-Based Alloys, *Oxidation of Metals*, **76**, 43-55, 2011

[9]V. Deodeshmukh and B. Gleeson, Evaluation of the hot corrosion resistance of commercial β -NiAl and developmental γ' -Ni3Al+ γ -Ni-based coatings, *Surface and Coatings Technology*, **202**, 643-647, 2007

[10]N. Eliaz, G. Shemesh, and R. Latanision, Hot corrosion in gas turbine components, *Engineering Failure Analysis*, **9**, 31-43, 2002

[11]Z. B. Bao, Q. M. Wang, W. Z. Li, et al, Preparation and hot corrosion behaviour of an Al-gradient NiCoCrAlYSiB coating on a Ni-base superalloy, *Corrosion Science*, **51**, 860-867, 2009

[12]X. Ren and F. Wang, High-temperature oxidation and hot-corrosion behavior of a sputtered NiCrAlY coating with and without aluminizing, *Surface and Coatings Technology*, **201**, 30-37, 2006

[13]M. Guo, Q. Wang, P. Ke, et al, The preparation and hot corrosion resistance of gradient NiCoCrAlYSiB coatings, *Surface and Coatings Technology*, **200**, 3942-3949, 2006

[14]P. D. Patil, V. G. Gude, A. Mannarswamy, et al, Optimization of direct conversion of wet algae to biodiesel under supercritical methanol conditions, *Bioresource Technology*, **102**, 118-122, 2011

[15]P. D. Patil, V. G. Gude, A. Mannarswamy, et al, Optimization of microwave-assisted transesterification of dry algal biomass using response surface methodology, *Bioresource Technology*, **102**, 1399-1405, 2011
[16]D. Lindberg, R. Backman, and P. Chartrand, Thermodynamic evaluation and optimization of the (Na2SO4+K2SO4+Na2S2O7+K2S2O7) system, *The Journal of Chemical Thermodynamics*, **38**, 1568-1583, 2006
[17]J. Goebel and F. Pettit, Na2SO4-induced accelerated oxidation (hot corrosion) of nickel, *Metallurgical and Materials Transactions B*, **1**, 1943-1954, 1970
[18]J. Goebel, F. Pettit, and G. Goward, Mechanisms for the hot corrosion of nickel-base alloys, *Metallurgical and Materials Transactions B*, **4**, 261-278, 1973

ORIGINAL IN-SITU METHOD TO QUANTIFY THE SIC ACTIVE CORROSION RATE AND ACTIVE/PASSIVE TRANSITION IN Ar/O_2 AND Ar/H_2O GAS MIXTURES AT VERY HIGH TEMPERATURES

Mathieu Q. Brisebourg, Bernard Boilait, Francis Rebillat, Francis Teyssandier

ABSTRACT

Oxidation behavior of silicon carbide under Ar/O_2 and Ar/H_2O gas mixtures was investigated at very high temperatures (>1700°C) in order to further understand both active oxidation kinetics as well as transition between active and passive oxidation modes. Experiments were conducted by Joule heating TEXTRON SCS-6 fiber samples inside an airtight vessel under different Ar/O_2 or Ar/H_2O environments. Temperature was regulated by means of a computer-assisted control loop consisting of a two-color pyrometer and a power supply delivering electrical current through the sample. Recordings of the variation of the electrical data relative to Joule heating were used to deduce in-situ information on the samples. All experimental results were compared to theoretical results obtained with a 3D finite volume simulation of transport phenomena inside the vessel. Active oxidation rates were found to increase linearly with P_{O2}, and to be independent of gas flow rate and of temperature up to 2100°C. Experimental results were in good agreement with simulation results and suggested that oxidation kinetics are limited by species transport through a free convection boundary layer. Active/passive transition results show good agreement with results from the literature obtained at lower temperature.

INTRODUCTION

Silicon Carbide (SiC) is a promising material for environmental barrier coating applications, due to its ability to form a continuous protective oxide scale (SiO_2) under highly oxidative environments (passive oxidation). However, at very high temperatures or less oxidative environments, SiC oxidation can lead to the formation of a gaseous product SiO(g), according to the following reactions :

$$O_2(g) + SiC(s) \rightarrow SiO(g) + CO(g) \quad (1)$$

$$2H_2O(g) + SiC(s) \rightarrow SiO(g) + CO(g) + 2H_2(g) \quad (2)$$

This oxidation mode is called "active oxidation" and results on a rapid degradation of the original material. Numerous studies have been conducted in the past in order to further understand the corrosion behavior of SiC under oxidative environments up to 1500°C[1,4-6], however its behavior at higher temperatures is still not precisely known as there are only few existing experimental methods able to provide an efficient in-situ approach of these phenomena over 1500°C. The main purposes of the present work are to develop and validate such an experimental apparatus, to study the influence of various physical and chemical parameters on the active corrosion of SiC under Ar/O_2 and Ar/H_2O gas mixtures, and to further understand the transition between the active and the passive mode of corrosion at very high temperatures. All experimental results were compared to theoretical results obtained with a 3D simulation of transport phenomena and solid/gas thermodynamic equilibria inside the corrosion furnace.

EXPERIMENTAL METHODS

Description of the experimental device

A schematic of the experimental device is presented in figure 1. The airtight vessel is provided with a gas inlet, a gas outlet as well as two copper electrodes. A conductive carbon support is attached to each of these electrodes. Each end of the sample is pasted to a carbon support using silver paste. A voltage can be applied between the copper electrodes using an external DC power supply in order to heat by Joule effect the sample to very high temperatures within a few seconds. The upper part of the vessel is provided with a quartz glass window so that the sample surface temperature can be measured using a two-color pyrometer located above the vessel. A homemade LABVIEW virtual instrument monitors both the pyrometer and the power supply in order to regulate the sample surface temperature and record all thermal and electrical data relative to a test.

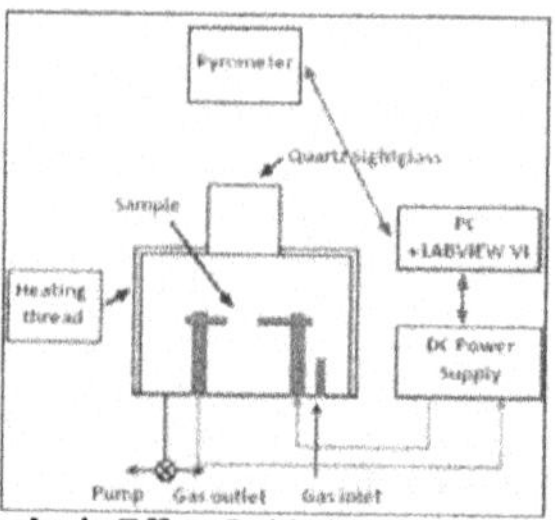

Figure 1. Schematic of the Joule Effect Oxidation Furnace experimental device

The gas mixture is prepared by means of a gas panel equipped with argon and oxygen gas flow controllers as well as water liquid flow controllers and vaporizer designed to generate controlled Ar/O_2 or Ar/H_2O atmospheres. Each part of the gas system downstream of the gas vaporizer (including the vessel's external wall) is wrapped in an electric heating thread so as to avoid water condensation.

Samples

The samples used in this study are 3.5 cm long TEXTRON SCS-6 fibers, the gauge length being 2.6 cm. The TEXTRON SCS-6 fiber is composed of[2]:

- an inner carbon core (33µm diameter), which is the conductive part of the sample where the heating power is generated by Joule effect
- a 1.5µm-thick pyrolytic carbon coating
- several two-phased SiC+C layers
- a 27µm-thick almost pure β-SiC layer which is the subject of this corrosion study.
- a 3µm-thick complex SiC/C outer-coating which is removed prior to the experiment by a 5h long treatment at 700°C under air and a 15 min long etching in a H_2O/HF 10% solution (Fig. 2).

During a test, the outer β-SiC layer is heated by the inner carbon core via radial heat conduction and reacts with the gaseous environments at the sample's outer surface.

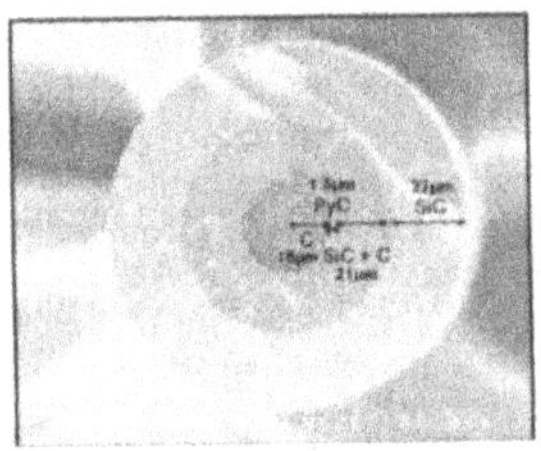

Figure 2. Scanning electron microscopy view of the cross-section of the TEXTRON SCS-6 sample after removal of the SiC/C outer-coating

Experimental procedure:

A pre-treated sample is pasted to the carbon supports using conductive silver paste. Primary vacuum is then maintained inside the vessel for 45 min. The gas mixture of the desired composition and at the desired flow rate is then introduced into the vessel. Sample is then heated to the set-point temperature within 20 sec. Sample surface temperature is maintained constant under flowing gas for 30 min or until failure of the fiber.

In-situ analysis of SiC corrosion using the electrical intensity recordings:

The electrical power generated in the carbon core of the sample (P_{fiber} in watts) by the Joule effect follows Joule's law:

$$P_{fiber} = R_{fibre} I^2 \tag{3}$$

where R_{fiber} is the fiber's electrical resistance in Ω and I is the electrical intensity. Assuming that a steady-state is almost instantly reached for all thermal transfers involving the sample and its surface, and that there is no energy loss inside the sample itself, conservation of energy implies that P_{fiber} is equal to the sum of the power losses due to the different heat transfer modes.

$$P_{rad} + P_{cond} + P_{conv} = P_{fiber} = R_{fiber}\, I^2 \tag{4}$$

where P_{rad} and P_{conv} are the power losses in watts due to radiative and convective heat transfer at the sample surface respectively and P_{cond} is the power loss in watts due to heat conduction towards the carbon supports, which is neglected in comparison with the two others. Using both Stefan's law and Newton's law of cooling, equation 4 can be rewritten as:

$$I^2 = \frac{2\pi r l\,(\sigma\varepsilon(T^4 - T_0^4) + h(T - T_0))}{R_{fiber}} \tag{5}$$

where r and l are the radius and the length of the sample (in m) respectively, σ is Stefan's constant in $W.m^{-2}.K^{-4}$, ε is the emissivity factor, h is the convective heat transfer coefficient in $W.m^{-2}.K^{-1}$ and T is the sample surface temperature in K.

As long as the sample's carbon core is not damaged by corrosion, R_{fiber} remains constant. If T is maintained at a constant value throughout the test, then I^2 is proportional to the sample's radius. When the sample's β-SiC outer layer is subjected to degradation, in-situ monitoring of its thickness throughout the test is thus possible via recording of the square of the electrical intensity.

The linear relationship between r and I^2 was experimentally verified by carrying out several oxidation tests at 1800°C under an Ar/O_2 mixture, P_{O2}=300Pa, and gas flow rate D = 100 SCCM for different lengths of oxidation. Each of these tests was interrupted as soon as the electrical intensity had decreased to a certain value. Figure 3 represents the variation of the square of the electrical intensity measured throughout the test. Cross-sections of the oxidized sample were then observed using scanning electron microscopy (SEM) in order to determine their radius. Figure 4 represents the square of the electrical intensity recorded at the end of each test as a function of the measured radii and provides experimental verification of the linear relationship between r and I^2.

A small intensity peak is observed at $I^2 \sim 0.04$ A^2. Though the origin of this peak is not yet precisely understood, SEM observations revealed that its occurrence coincides with the whole consumption of the 27μm-thick β-SiC outer layer.

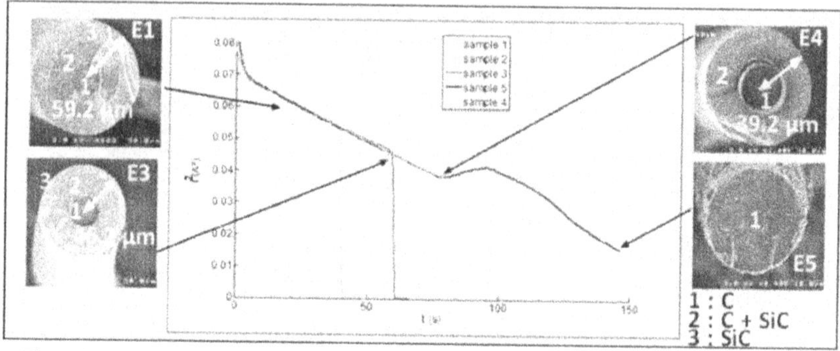

Figure 3. Variation of the square of the electrical intensity during oxidation of a TEXTRON SCS-6 fiber at 1800°C under Ar/O_2 at atmospheric pressure, P_{O2}=300 Pa

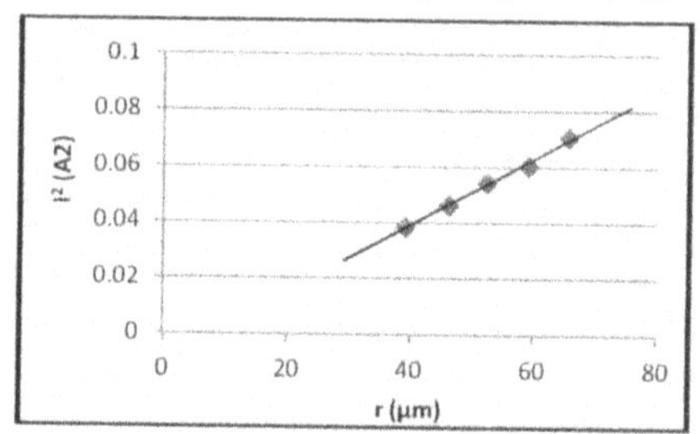

Figure 4. Square of electrical intensity as a function of sample's total radius from the oxidation tests described in figure 3.

In-situ analysis of SiC degradation using the electrical resistance recordings:

The electrical resistance measured during a test is the sum of the sample's resistance and the resistance of the rest of the system, including the sample/support resistance contacts which vary from a test to another. As a result, the electrical resistance recordings cannot provide in-situ quantitative information on the sample's corrosion. Nevertheless, they provide qualitative information on the state of the inner carbon core throughout the test.

Figure 5 represents the variation of the electrical resistance during the five tests conducted at 1800°C under Ar/O_2 gas mixtures at $P_{O2} = 300$ Pa. The value of resistance shows a significant increase after 100s oxidation test. SEM observation of sample 5 showed that this increase is related to the oxidation of the carbon core which damages the conductive properties of the sample. The time needed to consume entirely SCS-6 outer layers can thus be visualized through the electrical resistance recordings.

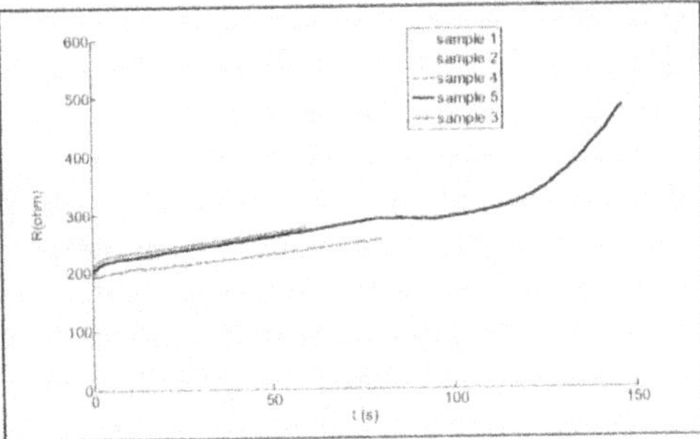

Figure 5. Variation of the square of electrical resistance during oxidation tests at 1800°C under Ar/O_2, P_{O2}=300 Pa

3D-SIMULATION OF TRANSPORT PHENOMENA INSIDE THE EXPERIMENTAL SYSTEM

3-D simulation of transport phenomena inside the experimental system by finite volume method was performed using the commercial solver ANSYS FLUENT 12. The Navier-Stokes equation, the heat equation as well as the Stefan-Maxwell equation for multi-component diffusion were solved in the gas phase domain. We assumed that:

- Steady-state was instantly reached for all transport phenomena
- Flow was laminar
- Heat generated inside the sample by Joule-effect was an homogeneous heat source
- Homogeneous and Heterogeneous thermodynamic equilibrium was reached at each point of the gas phase domain, including the gas/sample interface.
- Reactions occurring at the gas/sample interface has no influence on the gas composition at the outer boundary of the gas phase domain.

The 3-D graphical representation of the system as a well as a summarization of the boundary conditions applied on the different boundaries of the domain are shown in figure 6. The input parameters are the gas velocity and composition at the gas inlet, as well as the Joule-heating power inside the sample. Theoretical values for the SiC corrosion rate and gas composition at the SiC/gas interface as a function of the input parameters are derived from calculation results. The fact that formation of SiO_2 is thermodynamically allowed is deduced from the gas composition at the SiC/gas interface using Wagner's thermodynamic criterion for the active/passive transition[3]. SiO_2 is expected to form when oxygen activity at the surface is high enough for reaction 6 or reaction 7 to take place.

$$\frac{3}{2}O_2(g) + SiC(s) \rightarrow SiO_2(c) + CO(g) \tag{6}$$

$$3H_2O(g) + SiC(s) \rightarrow SiO_2(c) + CO(g) + 3H_2(g) \tag{7}$$

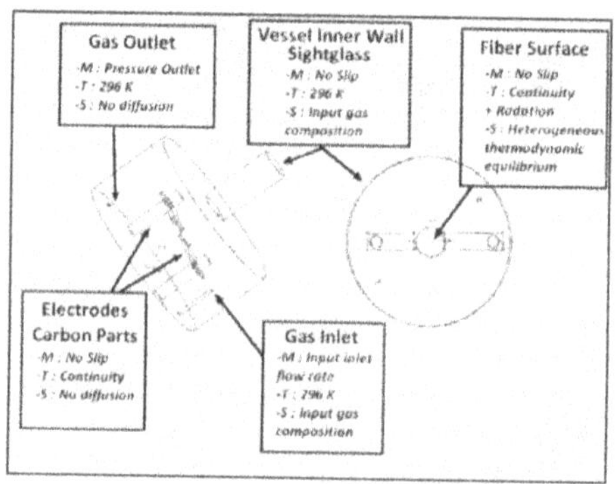

Figure 6. 3-D representation of the experimental device. Boundary conditions are mentioned for each kind of transport phenomena ("M", "T" and "S" for momentum, thermal and species transport respectively)

EXPERIMENTAL RESULTS

Active oxidation under Ar/O_2 atmosphere:

Figure 7 shows the variation of the square of the electrical intensity recorded during oxidation tests at T=1800°C and P=1 atm under Ar/O_2 gas mixtures with various oxygen partial pressures and a total flow rate of 100 sccm. The sample's total radius decreased linearly with time in all the studied conditions. SiC active oxidation rate was found to be proportional to P_{O2}. Figure 8 shows the variation of the square of the electrical intensity recorded during oxidation tests at T=1800°C and P=1atm under Ar/O_2 gas mixtures with P_{O2}= 300 Pa and various total flow rates. No clear dependence of the gas flow rate on the active oxidation kinetics was observed, except a slight increase of the corrosion rate for the maximum flow rate allowed by the gas panel (2000 sccm).

Temperature was also found to have no influence on SiC corrosion rate from 1700°C to 2100°C as shown in Figure 9. At 2100°C and 2200°C, however, the corrosion rate increased significantly. Electrical resistance recordings showed that the fiber's inner carbon core was indeed more rapidly subjected to oxidation at 2100°C and 2200°C that at lower temperatures, leading to a rapid increase of the fiber's electrical resistance.

Active oxidation under Ar/H_2O

Figure 10 shows the variation of the square of the electrical intensity recorded during corrosion tests at T=1800°C and P=1 atm under Ar/H_2O gas mixtures flowing at 1000 sccm and for various water vapor partial pressures. The sample's total radius decreased linearly with time in all the studied conditions. SiC active oxidation rate was found to be proportional to P_{H2O}. SiC degradation rate showed no dependence on temperature below 2100°C and in contrast, a significant increase at higher temperatures, as it was observed under Ar/O_2 mixtures.

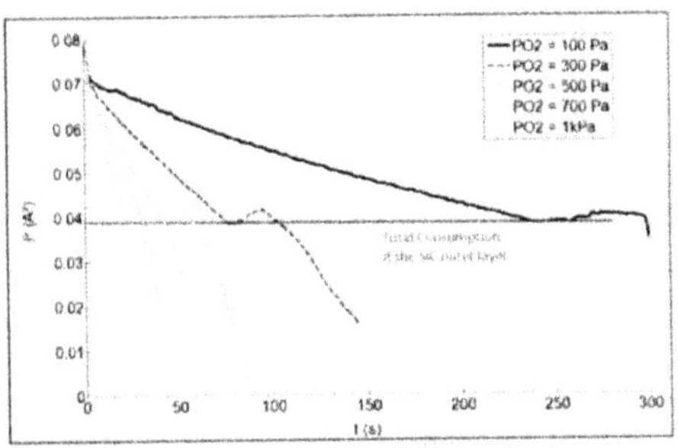

Figure 7. Variation of the square of electrical intensity during oxidation tests at 1800°C under Ar/O_2 at atmospheric pressure and various P_{O2}, Q=100 sccm

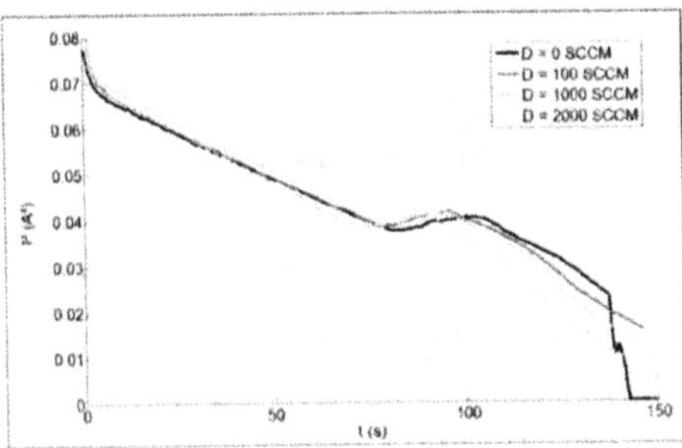

Figure 8. Variation of the square of electrical intensity during oxidation tests t 1800°C Ar/O_2 P_{O2}=300 Pa at various total gas flow rates (D)

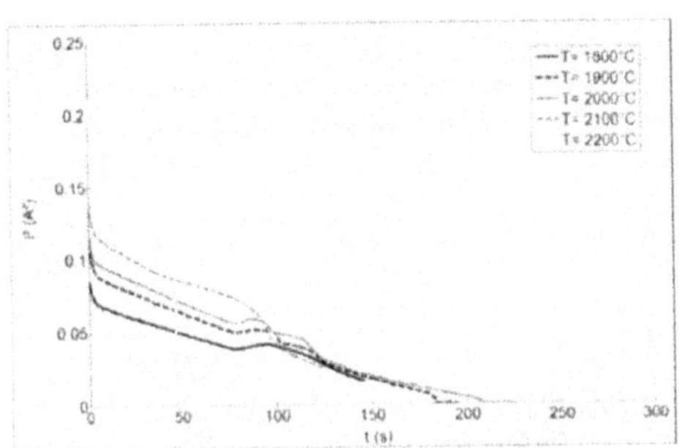

Figure 9. Variation of the square of electrical intensity during oxidation tests under Ar/O_2 P_{O2}=300 Pa at various temperatures

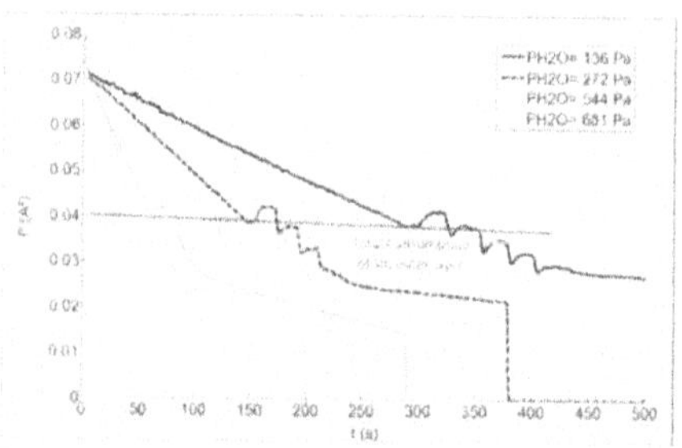

Figure 10. Variation of the square of the electrical intensity during oxidation tests at 1800°C under Ar/H_2O flowing at 1000 SCCM under various P_{H2O}

Active/passive transition under Ar/O_2 and Ar/H_2O:

Three different kinds of I^2 curves were observed during 30 min tests, ran at 1700°C under various atmospheres (Ar/O_2 and Ar/H_2O atmospheres with increasing oxidant partial pressure P_{OX}):

- At low oxidant partial pressures, I^2 decreased linearly in a similar manner than for the typical active corrosion curves previously represented in figures 7-10. Samples were entirely consumed before the end of the test.
- At high oxidant partial pressures, I^2 remained globally constant throughout the test. No significant radius variation was measured on the aged samples, and continuous oxide scales were identified on their surface, as shown on figure 11.b). This behavior is characteristic of a passive corrosion mode. Typical I^2 curves for the passive regime occasionally showed small peaks which are due to a temperature misreading induced by the oxide scale growth. Samples corroded in the passive mode occasionally experienced local failure before the end of the test.
- At intermediate oxidant partial pressures, I^2 tended to remain constant for a short duration before showing an abrupt decrease as the result of a fast corrosion of the sample prior to the end of the 30-min test. Small traces of discontinuous silica scales could sometimes be observed on the sample's surface, as shown on figure 11.a)

An example of each of these three kinds of I^2 curves observed at 1700°C under Ar/O_2 atmospheres is shown on figure 12. Many tests were conducted at various temperatures and P_{OX} in order to determine the fields corresponding to the different corrosion behaviors on a [10 000/T, log($P_{OX,}$)] map.

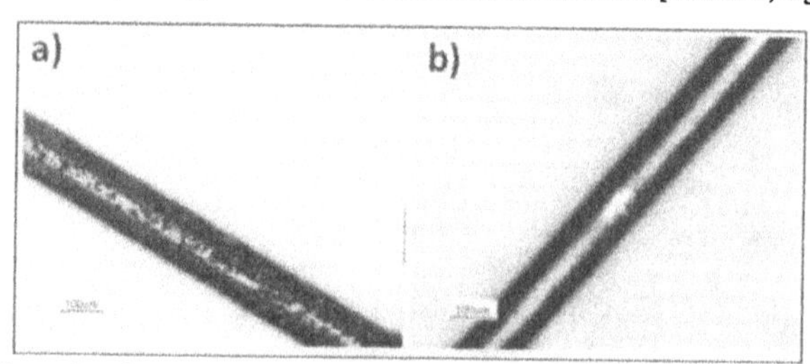

Figure 11. SiC samples after a corrosion test at 1700°C under a Ar/O_2 gas mixture flowing at 1000 SCCM a) at P_{O2} = 400 Pa, b) at P_{O2} = 800 Pa

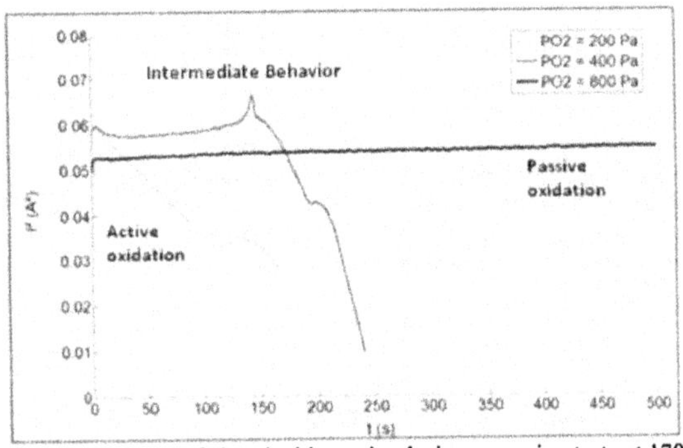

Figure 12. Variation of the square of electrical intensity during corrosion tests at 1700°C under Ar/O_2 flowing at 1000 SCCM under various P_{O2}

The active/passive transition boundary was defined for each temperature as the mean value of the highest oxidant partial pressure for which an active (or intermediate) behavior was observed, and the lowest oxidant partial pressure for which a passive behavior was observed.

DISCUSSION

Active oxidation kinetics:

SiC corrosion rate r_{deg} measured during active corrosion tests under Ar/O_2 and Ar/H_2O showed very good agreement with the theoretical values derived from the simulation results, as shown on figure 13. In agreement with the experimental results, the model shows no variation of the corrosion rate as a function of gas flow rate at temperatures below 2100°C.

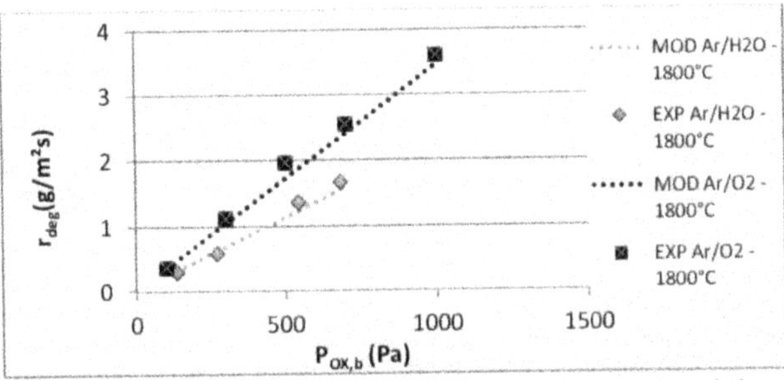

Figure 13. Experimental (EXP) and theoretical (MOD) values for SiC mass degradation obtained under Ar/O_2 and Ar/H_2O mixtures at 1800°C as a function of the oxidant partial pressure in the gas bulk

Hence, active corrosion kinetics under Ar/O_2 and Ar/H_2O below 2100°C appears to be rate-controlled by oxidant species diffusion through the gas phase . In such a case, r_{deg} can be simply written as:

$$r_{deg} = \frac{M_{SiC}\, h_{OX} \left(\frac{P_{OX,b}}{T_b} - \frac{P_{OX,s}}{T_s}\right)}{v_{OX} R} \tag{8}$$

Where M_{SiC} is the molar mass of SiC in g.mol^{-1}, R is the ideal gas constant in J.K^{-1}.mol^{-1}, h_{OX} is the mass transfer coefficient for the oxidant species diffusion through boundary layer in m.s^{-1}, v_{OX} is the stoichiometric coefficient for the oxidant species in the active corrosion reaction equation, $P_{OX,b}$ and $P_{OX,s}$ are the oxidant partial pressures in Pa in the gas bulk and at the SiC/gas interface respectively and T_b and T_s are the temperatures in K in the gas bulk and at the interface respectively. Simulation results for the gas composition at the SiC/gas interface showed that $P_{OX,s}$ is close to zero :

$$r_{deg} = \frac{M_{SiC}\, h_{OX}\, P_{OX,b}}{v_{OX} R T_b} \tag{9}$$

Equation 9 reveals the linear relationship between r_{deg} and $P_{OX,b}$. The lack of dependence of r_{deg} on gas flow rate under 2000 sccm can be explained by the fact that gas flow inside the experimental vessel is driven by free convection and gas phase dynamics at the inlet has no significant effect on it . h_{OX} also seems to be independent of T_s . h_{O2} and h_{H2O} were determined to be 0.23 m.s^{-1} and 0.30 m.s^{-1} respectively.

The model failed to predict the significant increase of r_{deg} above 2100°C, as can be seen in figure 14. This increase may be induced by sublimation of SiC which results in the formation of gases such as Si(g) and Si_2C(g)[4]. These latter are nevertheless not taken into account in the simulation.

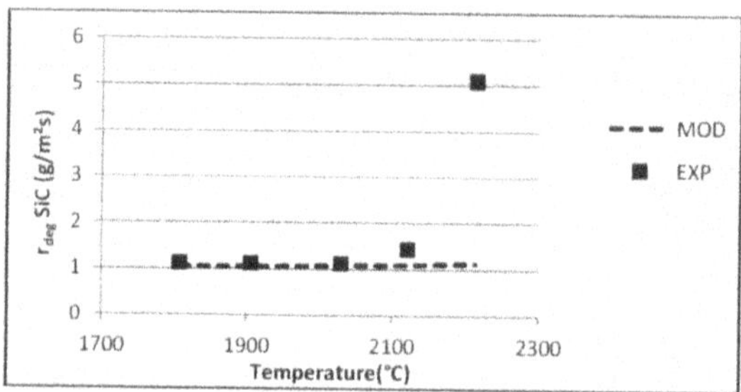

Figure 14. Experimental (EXP) and theoretical (MOD) values for SiC mass degradation rate obtained under Ar/O_2 mixtures at P_{O2}=300 Pa as a function of sample surface temperature

Active/Passive transition:

The active/passive transition measured in Ar/O_2 gas mixtures was found to be consistent with results from the literature obtained at lower temperatures with β-SiC as the studied material and argon as the carrier gas[4-6] (Fig. 15). Figure 16 shows that the active/passive transition measured in Ar/H_2O gas mixtures was consistent with results reported by Kim & Readey[7] although they were obtained for α-SiC under H_2/H_2O gas mixtures.

On the other hand, passive oxidation was experimentally observed under conditions for which 3-D simulation results predicted silica formation to be not allowed for thermodynamic reasons. The model used in this simulation widely overestimates the oxygen partial pressure in the bulk gas corresponding to the transition at a given temperature, as does Balat's analytical model[8] which is based on the same approach (Fig. 15).

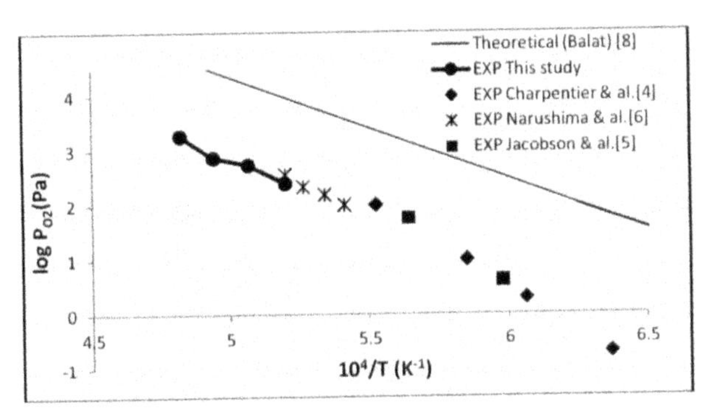

Figure 15. Experimental values for the active/passive transition of β-SiC under Ar/O2 gas mixtures compared with experimental and theoretical values extracted from the literature[4-6,8]

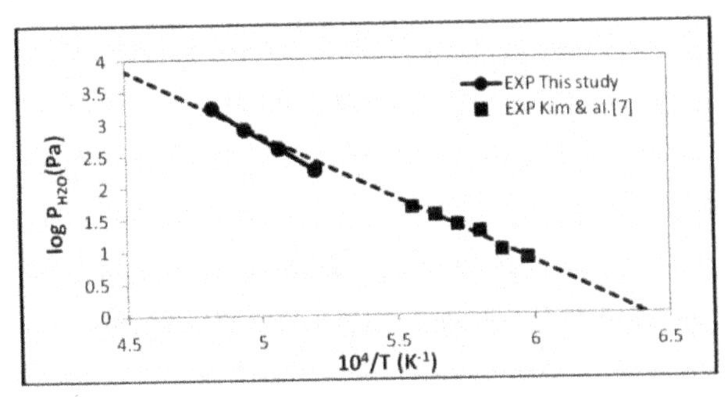

Figure 16. Experimental values for the active/passive transition of β-SiC under Ar/H_2O gas mixtures compared with experimental values extracted from the literature[7]

The intermediate corrosion behavior identified in this study presents some similarities with observations previously reported at lower temperatures by Jacobson & al[5] during TGA analysis of the break-down of an already-existing SiO_2 film on SiC. Indeed, some samples corroded under conditions belonging to the transition domain between pure active and pure passive corrosion showed a damaged silica scales on their surface, (figure 11.a)). Thus, active oxidation in this mode seems to be delayed by the slow degradation of an already-existing silica scale, which is likely formed during the 10-second heating step of the oxidation test.

CONCLUSION

A new experimental apparatus based on Joule-heating of SCS-6 fiber samples was designed to study active corrosion as well as active/passive transition of β-SiC under Ar/O_2 and Ar/H_2O

atmospheres at very high temperatures. Recordings of the square of the electrical intensity used to heat the sample by Joule effect allowed for an in-situ monitoring of the sample's total radius. Active oxidation kinetics were found to be rate-controlled by species diffusion through a free convection boundary layer, and independent of inlet gas velocity and of temperature up to 2100°C. SiC degradation rate showed a significant increase over 2100°C which was most likely caused by sublimation.

Active/passive transitions measured under Ar/O_2 and Ar/H_2O were consistent with results previously reported in the literature for lower temperatures.

A 3-D simulation of transport phenomena under thermodynamic equilibrium assumption inside the vessel provided very good prediction of SiC active degradation rates but failed to identify the active/passive transition under Ar/O_2 mixtures. A new model based on the study of the stability of an already-existing silica scale from a heterogeneous kinetics point of view is currently under study in order to bring forth a better description of the phenomena leading to this transition.

ACKNOWLEDGEMENTS

The authors would like to thank SNECMA Propulsion Solide and DGA for financial support, as well as Pr. G.L. Vignoles for his advice regarding the development of the 3-D model and Dr Zhu of NASA Glenn for reviewing this article.

REFERENCES

1 J.W. Hinze , H.C. Graham « The Active Oxidation of Si and SiC in the Viscous Gas-Flow Regime » J. Electrochem. Soc., 123 [7] 1066-1073 (1976)

2 X.J.Ning, P.Pirouz, « The microstructure of SCS-6 fiber » J. Mater. Res. 6 [11] (1991) 2234-48

3 C. Wagner « Passivity during the Oxidation of Silicona at Elevated Temperatures » J. Appl.Phys. Soc. 29 [9] 1295-97 (1958)

4 L.Charpentier, M.Balat-Pichelin, H.Glénat, E.Bêche, E.Laborde, F.Audubert, « High-Temperature oxidation of SiC under helium with low pressure oxygen – Part 2 : CVD β-SiC » J. Europ. Ceram. Soc. 30 (2010) 2661-2670

5 N.S. Jacobson, D.L. Myers « Active Oxidation of SiC» Oxyd. Met. 1-25 (2011)

6 T. Narushima, T. Goto, T. Hirai, Y. Iguchi « High-Temperature Active Oxidation of Chemically Vapor-Deposited Silicon Carbide in an Ar-O2 Atmosphere » J. Am. Ceram. Soc. 74 [10] 2583-86 (1991)

7 H-E. Kim, W. Readey. "Active Oxidation of SiC in low dew-point hydrogen above 1400°C" Silicon Carbide '87 pp 301-312 (1989)

8 M. Balat, G. Flamand, G.Male, G. Pichelin, « Active-to-Passive Transition in the Oxidation of Silicon Carbide at High-Temperature and Low Pressure in Molecular and Atomic Oxygen ». J. Mater. Science. 27 697-703 (1992)

SILICON CARBIDE NANOTUBE OXIDATION AT HIGH TEMPERATURES

Nadia Ahlborg * and Dongming Zhu †
*Department of Materials Science and Engineering, The Ohio State University, Columbus, Ohio, USA
†Durability and Protective Coatings Branch, NASA Glenn Research Center, Cleveland, Ohio, USA

ABSTRACT

Silicon Carbide Nanotubes (SiCNTs) have high mechanical strength and also possess many desirable functional properties in high temperature and extreme environments, thus becoming one of the most promising materials for multifunctional composite applications. In this study, SiCNTs were investigated for use in strengthening high temperature silicate and oxide materials for high performance ceramic nanocomposites and environmental barrier coating bond coats. The high temperature oxidation behavior of the nanotubes was of particular interest. The SiCNTs were synthesized by a direct reactive conversion process of multiwall carbon nanotubes and silicon at high temperature. Thermogravimetric analysis (TGA) was used to study the oxidation kinetics of SiCNTs at temperatures ranging from 800 to 1300°C. The specific oxidation mechanisms were also investigated. The stability of silicon-SiCNT bond coats with an $Yb_2Si_2O_7$ silicate top coat was also examined.

1. INTRODUCTION

Silicon Carbide (SiC) has excellent mechanical and thermal properties, which makes it a desirable material in high temperature applications. Silicon Carbide Nanotubes (SiCNTs) and their composites potentially have combined high strength, high toughness, and improved multifunctional properties. Therefore, they are currently being considered for composite environmental barrier coatings and bond coats for protecting ceramic matrix composites (CMCs) toward improving the coating overall high temperature durability and performance [1]. However, because SiCNTs are nano-size and tube-like fibers, their stability in oxidizing matrices such as oxides and silicates is of a major concern. In this study, the SiCNTs were investigated for their viability as a reinforcement material in high stability oxide-based environmental barrier coatings. Because such coatings are exposed to high temperatures and harsh conditions, an understanding of oxidation kinetics of the SiCNTs is important. This work was undertaken to determine the activation energy of SiCNT oxidation in pure O_2 and determine an upper use temperature. The stability of the SiCNTs in a silicon and ytterbium silicate coating system was also compared with that of the unprotected SiCNTs in the oxidation kinetics studies.

2. EXPERIMENTAL METHODS

2.1 Synthesis of SiCNTs

SiCNTs were synthesized by a direct reactive conversion process of multiwall carbon nanotubes and silicon. The silicon and nanotubes powders were placed in physical contact and heat treated for eight hours at 1400°C. Heat treatments were carried out in an Ar+5%H_2 reducing environment using a Carbolite tube furnace. The processing optimization of the SiCNTs has been discussed elsewhere [2, 3].

2.2 TGA experiments

Thermogravimetric analysis (TGA) experiments were carried out by oxidizing a small amount of SiCNT powder (approximately 0.2 grams) in an alumina cup container using Cahn 1000 microbalance in the temperature range of 800 to 1300°C. Each test was carried out isothermally at a given temperature in 100% dry O_2 for 100 hours. Following heat treatment, the sample was further analyzed using scanning electron microscopy (SEM) and x-ray diffraction (XRD) for morphology and phase characterization.

2.3 Composite coating experiments

Two specimens consisting of a Si-SiCNT and $Yb_2Si_2O_7$ layered coating on melt-infiltrated (MI) SiC/SiC CMC substrate were produced and examined. The substrate specimens were 25 mm in diameter and 2 mm thick. The SiCNT-Si bond coats were made by processing a mixture of Si and SiCNT powders that were placed on top of SiC-SiC CMC at 1410°C in Ar+5%H_2. $Yb_2Si_2O_7$ top coats were made using a hot press technique. Table 1 summarizes the composition and processing conditions for each specimen. The coatings were furnace tested in air at 1300°C for 50 hr to investigate the coating-nanotube system stability.

Table 1: Composite coating samples

	SAMPLE No. 1	SAMPLE No. 2
SUBSTRATE	GenII SiC-SiC CMC	GenII SiC-SiC CMC
BOND COAT	SiCNT + 50wt%Si 1410°C in Ar + 5%H_2	SiCNT 1410°C in Ar+H_2
TOP COAT	$Yb_2Si_2O_7$, hot pressed, 1.2 mm thick	$Yb_2Si_2O_7$, hot pressed, 1.2 mm thick
OTHER TREATMENTS	n/a	heat treatment, 50 hr, 1300°C air

3. RESULTS AND DISCUSSION

3.1 Rate constant and activation energy

Previous work predicted that the oxidation of SiC follows a parabolic rate law [4-6],

$$\left(\frac{\Delta m}{SA}\right)^2 = K_p^{\bullet} t + C \tag{1}$$

where $\Delta m/SA$ is the specific weight gain of the specimen after oxidation, $K_p^{\bullet}$ is the constant pressure oxidation parabolic rate constant, t is time, and C is a constant. Figure 1 shows the specific weight changes plotted as a function of time. The oxidation behavior appears pseudo-parabolic at

800°C up to 1300°C for fiber-like SiCNTs, and K_p^* was calculated using Equation (1). For 1000°C to 1300°C, shrinking of the fiber surface area causes the reaction to slow down over time. For these temperatures, only the initial, constant-slope portion of each curve was used to calculate K_p^*. These values for K_p^* are shown in Table 2.

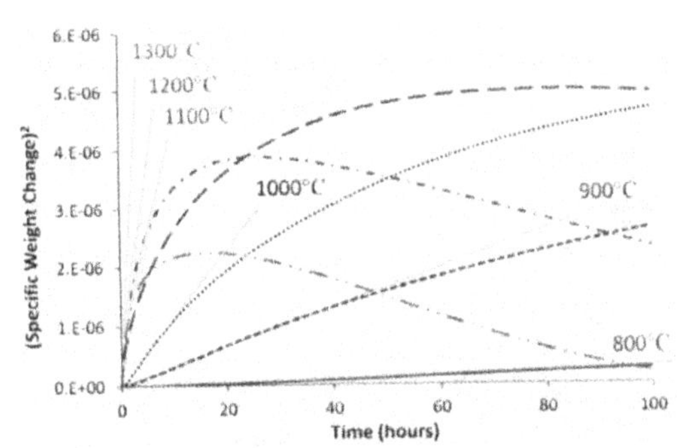

Figure 1: Specific weight changes of SiCNTs as a function of time. The slope of each curve gives the rate constant $\mathbf{K_p^*}$.

Table 2: Rate constant values (in O_2 unless indicated)

TEMPERATURE (°C)	AVERAGE RATE CONSTANT (mg^2-cm^{-4}-h^{-1})
800	3.44×10^{-9}
900	2.81×10^{-8}
1000	7.93×10^{-8}
1100	2.47×10^{-7}
1200	5.95×10^{-7}
1300	8.82×10^{-7}
1000 in air	2.60×10^{-8}

The rate constants (K_p^*) are related to temperature by the Arrhenius equation:

$$K_p^* = A\exp(-Q/RT) \quad (2a)$$

$$\ln K_p^* = \ln A - Q/RT \quad (2b)$$

where Q is the activation energy, R is the ideal constant, T is the temperature, and A is a constant. A graph of $\ln(K_p^*)$ vs 1/T gives a line with slope Q/R, as shown in Figure 2. Applying Equation (2) gives an activation energy of 155 kJ/mol. The activation energy is in good agreement with literature reported data [4-7], and especially for SiC oxidation by O_2 diffusion through vitreous silica [7].

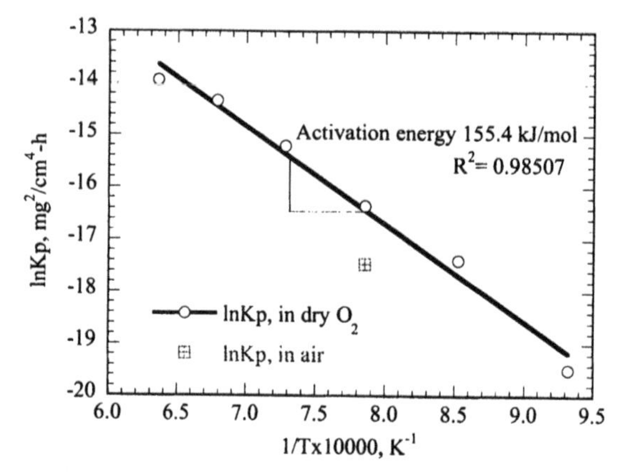

Figure 2: Arrhenius plot of the oxidation rate constants of SiCNT. The activation energy is calculated from the slope of the least squares linear fit.

3.2 X-ray analysis

X-ray analysis was performed to identify the material phases present in the nanotubes. Figure 3 summarizes this x-ray data. Carbon nanotubes present in the as-received powders burned off as the furnace initially heated up. SiO_2 is undetectable by XRD in the samples from the 800 to 1000°C tests, indicating the formation of an amorphous scale. Both crystalline SiO_2 and SiC are present in the samples tested at 1100°C and 1200°C. After 100 hours of testing at 1300°C, there is no detectible SiC remaining (SiCNTs fully oxidized), and the scales become almost fully crystalline phases.

3.3 SEM analysis

SEM images of as-processed and heat treated SiCNTs are shown in Figure 4. After 100 hours of heat treatment in O_2, a SiO_2 scale forms on the nanotube surface (Figure 4 a-d). The degree of SiO_2 growth was quantified by measuring the increase in the nanotube diameter over the 100 hour heat treatment period. The parabolic rate law can also be expressed using this diameter increase,

$$(\Delta d)^2 = K_P t + C \qquad (3)$$

Applying Equations (2) and (3) gives an activation energy of 142 kJ/mol, which is in good agreement with the TGA calculations, and the published literature values [7].

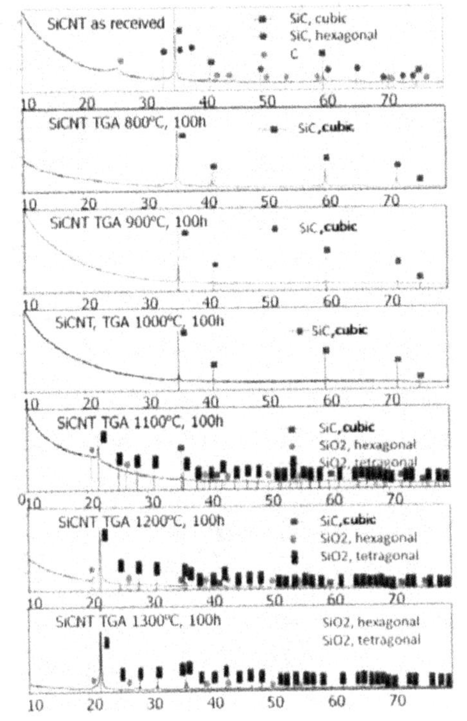

Figure 3: X-ray diffraction spectra for as-processed and heat treated SiCNTs.

After 100 hours of heat treatment at very high temperatures (1100 to 1300°C), nearly all of the initial SiC cores have reacted. This caused the nanotube structure to deteriorate (Figure 4 e and f). The changes in nanotube diameter as a function of temperature are illustrated in Figure 5.

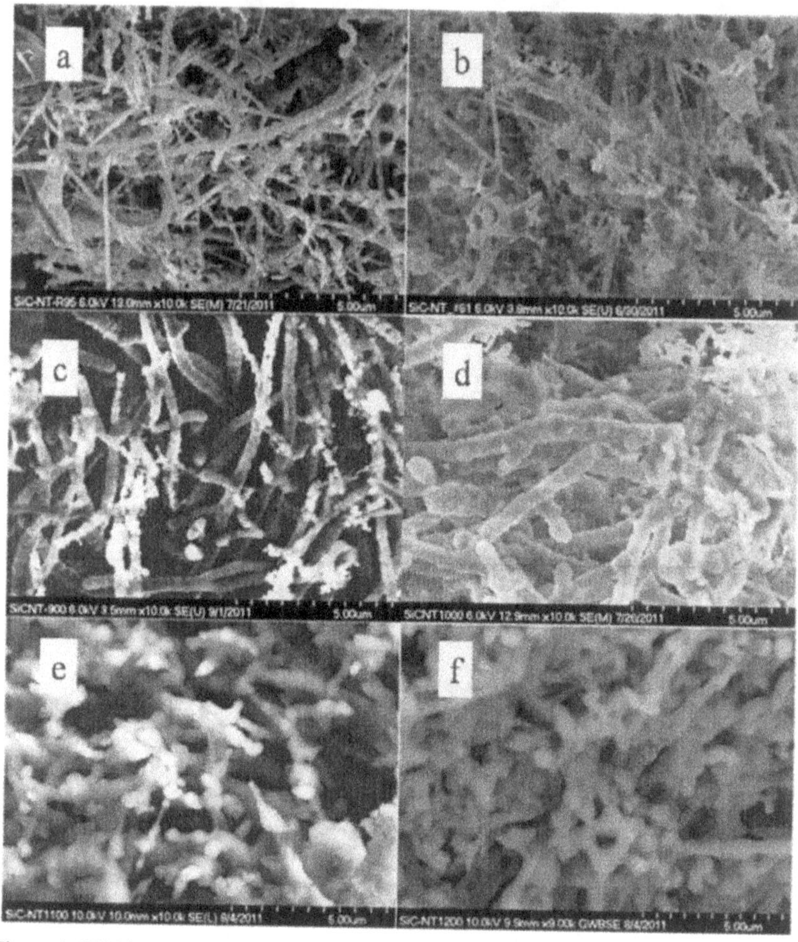

Figure 4: SEM images of SiCNTs (a) as-processed, and SiCNTs heat treated for 100 hours at (b) 800°C, (c) 900°C, (d) 1000°C, (e) 1100°C, and (f) 1200°C.

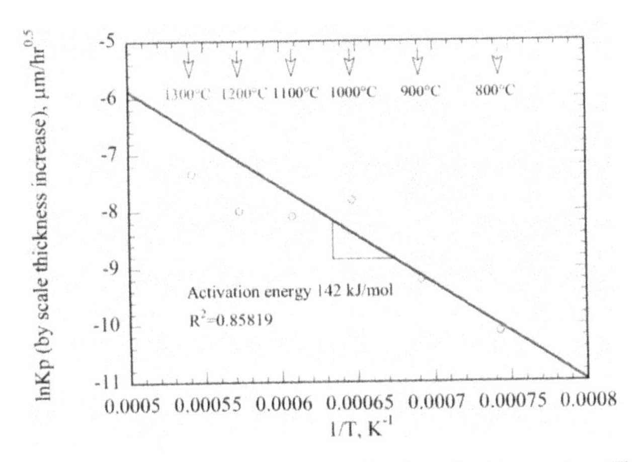

Figure 5: SiCNT diameter increase as a function of test temperature. The activation energy is given by the slope of the least squares linear fit.

3.4 SiCNTs with Si and $Yb_2Si_2O_7$ protective coating

SEM images were obtained of the cross-section of the two coated CMCs and are shown in Figure s 6 and 7. Nanotubes are present in both as processed and the furnace heat treated samples, indicating that the coating provides some amount of environmental protection for the SiCNTs.

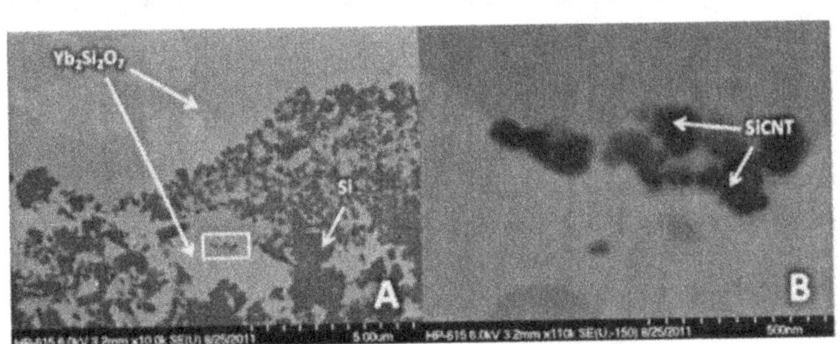

Figure 6: SEM images of sample # 1 (a) at the $Yb_2Si_2O_7$ Si-SiCNT interface and (b) SiCNTs embedded in $Yb_2Si_2O_7$.

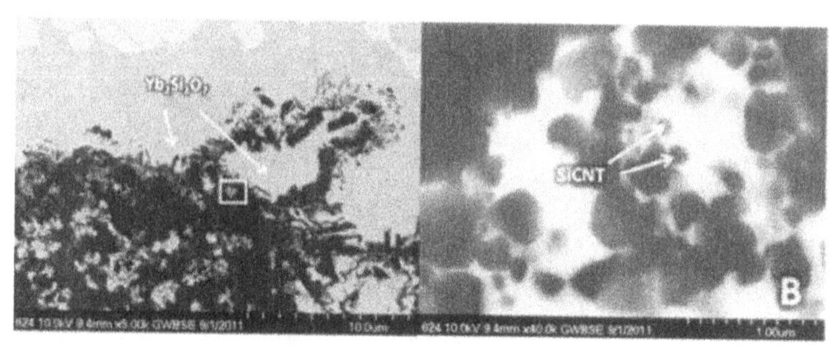

Figure 7: SEM images of sample #2 (a) at the $Yb_2Si_2O_7$ and Si-SiCNT interface and (b) SiCNTs embedded in $Yb_2Si_2O_7$.

4. SUMMARY AND CONCLUSION

Pseudo-parabolic oxidation behavior was observed for SiCNTs in the temperature range of 800 to 1300°C in dry O_2 due to the protective SiO_2 scale formation and reduced oxidation interface surface area of the fibrous structure nanotubes. Based on TGA and scale thickness kinetics studies, the activation energy was determined to be 142-155 kJ/mol. The activation energy is in good agreement with literature values for SiC oxidation with O_2 diffusion through vitreous silica. SiCNTs oxidized slowly and maintained relatively long-term stability in O_2 at temperatures of 800 to 1000°C, retaining their nanotube structure for 100 hours in pure O_2. The SiCNTs oxidized and breakdown rapidly at 1100°C and 1200°C, and the nanotube structure was completely destroyed within short period oxidation. Placing a protective coating of $Yb_2Si_2O_7$ over the nanotubes showed a dramatic reduction in damage to the SiCNT structure, and therefore may have the potential to provide protection from O_2 up to 1300°C.

ACKNOWLEDGMENTS

This research was made possible by the Lewis Educational and Research Collaborative Internship Project (LERCIP), the NASA Supersonics project, and the NASA Environmentally Responsibly Aviation project. The authors thank Donald Humphrey and Rahul Mittal at Glenn Research Center for their assistance with experiments and Durability and Protective Coatings Branch Chief Joyce A. Dever.

REFERENCES

[1] Yoshiki Yamada and Dongming Zhu, "Finite Element Model Characterization of Nano-Composite Thermal and Environmental Barrier Coatings", The 35th International Conference on Advanced Ceramics and Composites, January 2011.

[2] T. Taguchi, N. Igawa, and H. Yamamoto, "Synthesis of Silicon Carbide Nanotubes." J. Am. Ceram. Soc., 88(2), 459-461 (2005).

[3] Samantha Benkel and Dongming Zhu, "Phase Stability and Thermal Conductivity of Composite Environmental Barrier Coatings on SiC/SiC Ceramic Matrix Composites", The 35th International Conference on Advanced Ceramics & Composites, January 23-29, 2011.

[4] Krishan L. Luthra, "Some New Perspectives on Oxidation of Silicon Carbide and Silicon Nitride," J. Am. Ceram. Soc. 74 (5), 1095-1103 (1991).

[5] W. W. Pultz, "Oxidation of Submicroscopic Fibrous Silicon Carbide", J. Am. Ceram. Soc., 50(8), 419-420 (1967).

[6] Takayuki Narushima, Takashi Goto, and Toshio Hirai, "High-Temperature Passive Oxidation of Chemically Vapor Deposited Silicon Carbide", 72 [8] 1386-90 (1989).

[7] Volker Presser and Klaus G. Nickel, "Silica on Silicon Carbide", Critical Reviews in Solid State and Materials Sciences, 33 (1) 1–99 (2008).

Advanced Wear-Corrosion Resistant, Nano-Composite and Multi-Functional Coatings

EVALUATIONS OF MULTILAYER COATINGS FOR GALVANIC CORROSION RESISTANCE APPLICATIONS – OXIDE AND NITRIDE COATINGS ON CARBON STEEL

C. Qu,[1] R. Kasica,[2] R. Wei,[3] E. McCarty,[4] J. H. Fan,[1] D. D. Edwards,[1] G. Wynick,[1] L. Lin,[1] Y. Liu,[1] R. E. Miller,[1] and X. W. Wang[1*]

1. School of Engineering, Alfred University, Alfred, NY 14802, USA
2. Center for Nanoscale Science and Technology, NIST, Gaithersburg, MD 20899, USA
3. Southwest Research Institute, San Antonio, TX 78238, USA
4. Materials Technologies Consulting, LLC, Clarkston, MI 48346, USA

ABSTRACT

Magnesium alloy materials are utilized in some modern vehicle designs due to their low mass densities. The challenge is to mitigate the galvanic corrosion when a magnesium alloy part is fastened with a carbon steel bolt. Current solutions involve aluminum fasteners, isolators and sleeves. The aluminum parts can only partially isolate steel and steel fasteners from magnesium as the thickness of the natively grown aluminum oxide on the aluminum metal surface may vary from one spot to another. Additional aluminum surface treatment procedures may reduce the corrosion, but cannot eliminate the corrosion completely. This study will show that three layer coatings, when applied to steel parts, can electrically insulate the steel from magnesium. The layer stacking sequence is silicon nitride, aluminum oxide and UV curable aluminum oxide. Based on zero-resistance ammeter (ZRA) measurement and ASTM B 117 salt mist spray testing, the three layer coating mitigates the galvanic corrosion better than that provided by aluminum. Material properties and corrosion testing results will be provided.

INTRODUCTION

In order to reduce fuel consumption rates, light-weight materials are being utilized in modern vehicles.[1] Among all available metals for automotive applications, magnesium alloys are the lightest materials. Although aluminum fasteners are typically used to join magnesium in vehicle assemblies, steel fasteners are desired for their low cost, ductility, and durability. Since a galvanic cell exists between a magnesium alloy piece and a carbon steel piece, galvanic corrosion takes place.[2-3] Current solutions involve aluminum fasteners, isolators and sleeves. However, the natively grown aluminum oxide on the surface of aluminum can vary in thickness from one spot to another and, consequently, only partially isolate the steel from the magnesium. In the last four years, we experimented with ceramic thin film coatings on carbon steel substrates and systematically studied the isolation effects due to the thin film coatings.[4-9] In this paper, some of the earlier results are summarized, along with the new results. Several different types of ceramic thin films are fabricated, including aluminum oxide and silicon nitride. Aluminum oxide thin films are fabricated with e-beam evaporation technique; silicon nitride thin films are fabricated with PECVD. UV-curable ceramic embedded polymer coatings (cerium oxide, aluminum oxide or the mixture of the two) are also utilized as the top layer. UV-curable precursors[1] are $R(OCH_2CH_2OCH_3)_3)$ (R stands for Al and/or Ce) and a liquid reaction-initiator-containing polymer material (AR Base). The detailed information for the coating techniques is discussed in the previous papers. The films are analyzed by SEM and electrochemical measurements, including open circuit potential (OCP), polarization and impedance measurements.

[1] Purchased from Chemat, Inc.

EXPERIMENTS

Sample preparation

Two types of carbon steel substrate materials are utilized: flat 1050 carbon steel substrates for electrochemical measurements and Kamax M8 carbon steel bolts for salt spray measurements. Two types of magnesium alloy plates are utilized: AM60B for ZRA(zero resistance ammeter) measurements and Magnesium Elektron AZ31 for the salt spray measurements.

Substrates (carbon steel coupons and Mg plates) are polished with sandpapers with the grit of 120, 220, 400, 600 and 1200 in a sequence. They are then cleaned in an ultrasonic IPA bath and dried with compressed air.

Electrochemical measurements

Electrochemical measurements refer to EIS (electrochemical impedance spectroscopy) and ZRA (zero resistance ammeter) measurements. The EIS measurements include OCP (open circuit potential), polarization and impedance measurements. A Solartron 1260 analyzer and a Potentiostat/Galvanostat Solartron 1287 are utilized for EIS measurements. Before testing, a mask (purchased from Gamry Inc.) was placed on the samples to yield the normalized testing area of 1 cm^2. The platinum electrode (counter electrode) has the total surface area of 18 cm^2. The reference electrode of saturated calomel electron (SCE) with a reference potential of +0.241 V (vs. saturated hydrogen electron) was purchased from Fisher Scientific, Inc. Carbon steel plates (coated or uncoated) are utilized as the working electrode. See Figure 1.

Inside of the glass tube is 5 wt.% NaCl solution. The diameter of the glass tube is 2.5 cm and the water level is about 13 cm high from the sample (working electrode). The reference electrode is 3 cm below the water level. The distance between the platinum electrode (counter electrode) and the sample (working electrode) is about 2 cm.

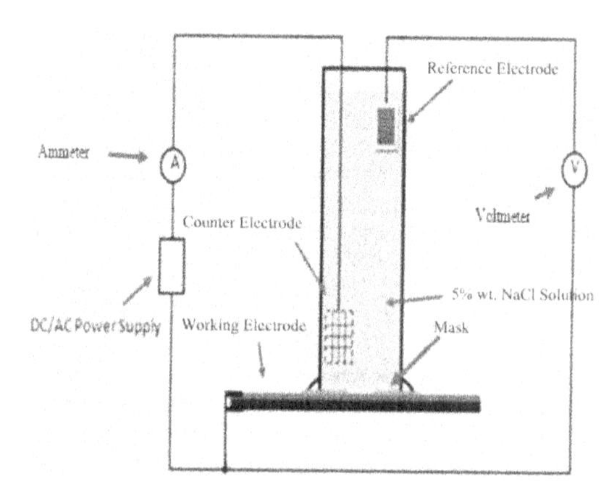

Figure 1 An illustration of the setup for EIS testing

In the zero resistance ammeter (ZRA) measurement illustrated in Figure 2, a magnesium plate covered with masking tapes is placed in a 5 wt.% salt water bath (on the right hand side), with an exposed area (towards the left side). The surface area exposed is approximately 1 cm^2. A steel plate with or without a coating is also placed in the bath (on the left hand side). With the masking tapes, the exposed area is also approximately 1 cm^2 (towards the right side).

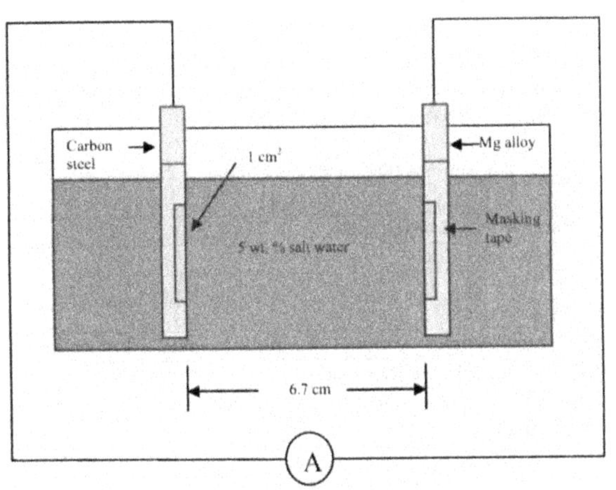

Figure 2 An illustration of ZRA (zero resistance ammeter) measurement set up

Salt mist spray tests

Tests using coated or uncoated bolts, attached to AZ31 magnesium plates from Magnesium Elektron, then exposed to a steady stream of salt mist (also 5 wt.% NaCl), are analyzed over a period of 3 days. The M8 bolt and the Mg plate were fastened onto a nylon plate with an M8 nut (See Figure 3).

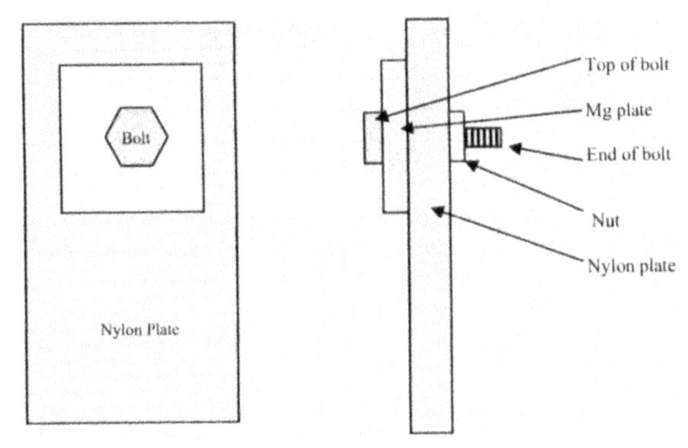

Figure 3 An illustration of bolt-Mg assembly

A commercially available corrosion test chamber (MPF-20 manufactured by Auto Technology) was utilized for the salt mist spray tests. Samples were tilted 45°, and rotated 90° on a daily base. Test conditions followed the ASTM B117 standard. Photos were taken each day during tests for visual inspections. Samples were weighed to 0.1 mg before and after testing (after the removal of corrosion products by sand-blasting).

RESULTS

The morphologies of aluminum oxide and silicon nitride coatings are provided in Figure 4. The cluster size of aluminum oxide is approximately 80 nm while that of silicon nitride is larger than 1 micrometer. In Table 1, a summary is given for samples tested with electrochemical measurements.

Table 1 Summary of sample IDs for electrochemical measurements

Sample ID	Coating
A	No coating, Bare 1050 carbon steel plate
B	As-deposited aluminum oxide
C	Silicon nitride
D_1	UV-curable cerium oxide
D_2	UV-curable aluminum oxide
E	Two-layer aluminum oxide with heat treatment
F	Combined coating of two-layer aluminum oxide with heat and UV-curable cerium oxide
G	Aluminum (Baseline)
H	Bottom: Silicon nitride; Top: Aluminum oxide
I	Bottom: Silicon nitride; Top: UV curable aluminum oxide
J	Bottom: Aluminum oxide; Top: UV curable aluminum oxide
K	Bottom: Silicon nitride; Middle: Aluminum oxide; Top: UV curable aluminum oxide

In Table 2, a summary of the impedance modulus of samples A, B, C and D_1 is provided for single layer coated substrates along with control.

Table 2 A summary of impedance mudulus at 0.1 Hz

Sample ID	Impedance Modulus at 0.1 Hz (Ω-cm^2)
A	1.7×10^3
B	2.4×10^4
C	8.5×10^6
D_1	1.1×10^5

In Figure 5, the open circuit potential (OCP) is plotted as a function of time. The middle-blue curve is for a typical aluminum oxide thin film coated on 1050 carbon steel, and the bottom-green curve is for uncoated 1050 carbon steel. In general, the OCP value of the aluminum oxide film is higher than that of the uncoated carbon steel sample.

Figure 4 Left: Top view of aluminum oxide film with 100K magnification (Sample B)
Right: Top view of silicon nitride film with 100K magnification (Sample C)

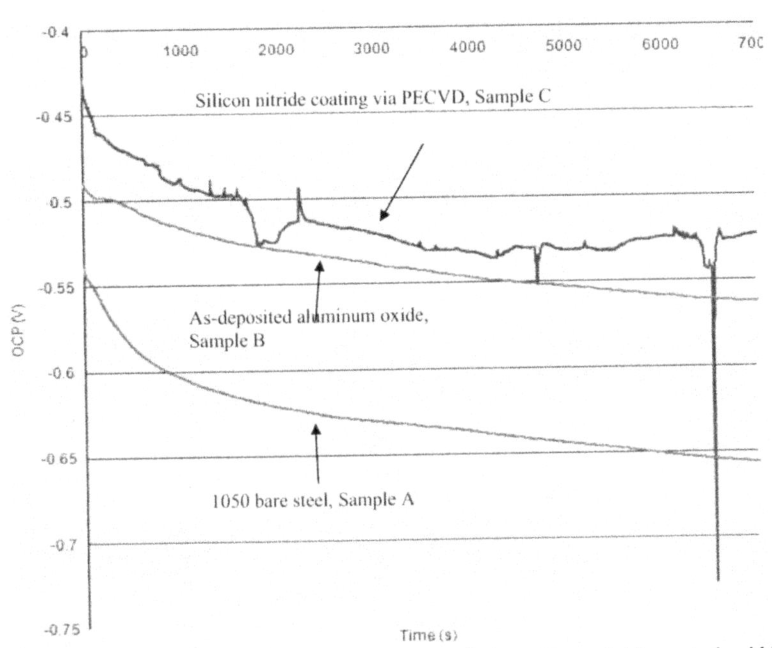

Figure 5 Open circuit potential vs. time. Top-red curve is for a silicon nitride on steel, middle-blue curve is for an aluminum oxide thin film on steel, and the bottom-green curve is for uncoated steel.

In Figure 6 three polarization curves are provided, with the green curve for uncoated 1050 carbon steel and the blue curve for aluminum oxide coated 1050 steel.

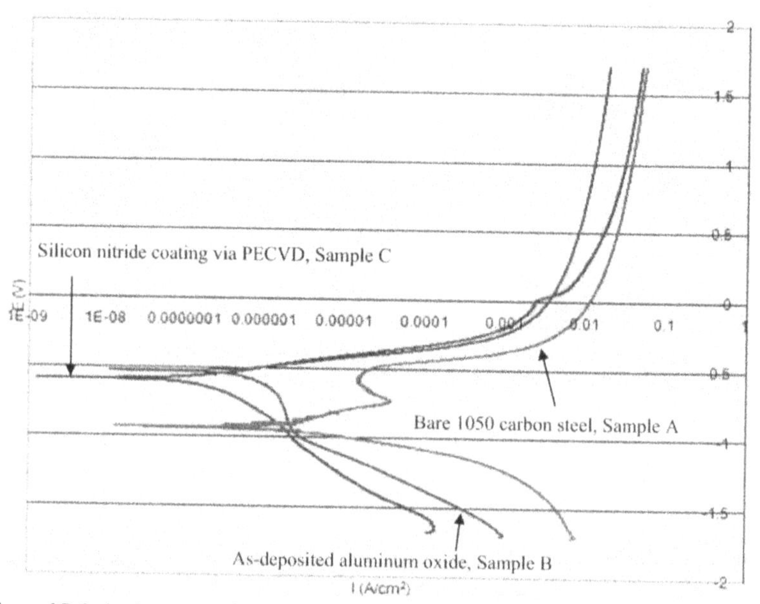

Figure 6 Polarization curves for uncoated steel substrate (green), aluminum oxide coating on steel (blue) and silicon nitride coating on steel (red)

Attempts were made to optimize the single layer aluminum oxide coating deposited on 1050 carbon steel by adjusting the film thickness, deposition rate and substrate temperature. Efforts were with with limited improvements and none of the single layer aluminum oxide coatings were able to match the performance of the silicon nitride ($8.5x10^6$ Ω-cm^2).

The next approach was to develop and evaluate two layer aluminum oxide coating in which the first layer (150 nm) is heat treated at 350°C for 40 minutes followed by the deposition of a second layer (120 nm) (Sample E). The additional step increases the impedance at 0.1 Hz to $3x10^5$ Ω-cm^2, which is almost one order of magnitude greater than the aluminum oxide sample without heat treatment (Sample B) with an impedance of $4x10^4$ Ω-cm^2. However, even with this additional heat treat step, the two layer aluminum oxide coating cannot match the impedance of silicon nitride ($8.5x10^6$ Ω-cm^2). For the impedance vs. frequency plot for each of the samples, see Figure 7.

In Figure 8, the open circuit potentials are provided for heat treated and as-deposited aluminum oxide coated carbon steel samples. The potential for the heat treated aluminum oxide shows the unstableness, a similar feature observed for silicon nitride coated steel sample. In Figure 9, the polarization currents are provided for heat treated and as-deposited aluminum oxide coated carbon steel samples. The corrosion current is much lower than that of the as-deposited sample.

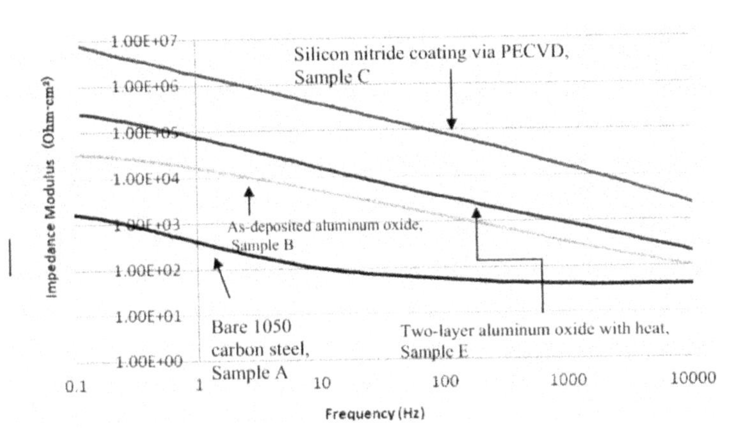

Figure 7 Impedance modulus vs. frequency
Top blue curve: silicon nitride coating;
Middle red curve: heat treated aluminum oxide;
Middle green curve: as-deposited aluminum oxide
Bottom black curve: bare 1050 carbon steel control sample

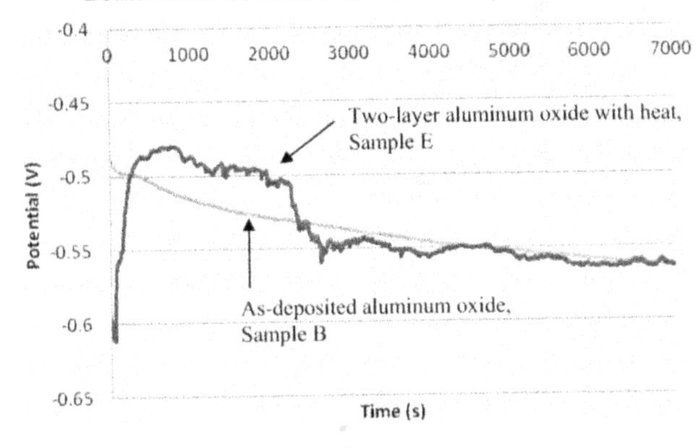

Figure 8 Open circuit potential vs. time
Red curve: two layer aluminum oxide with heat;
Green curve: as-deposited aluminum oxide

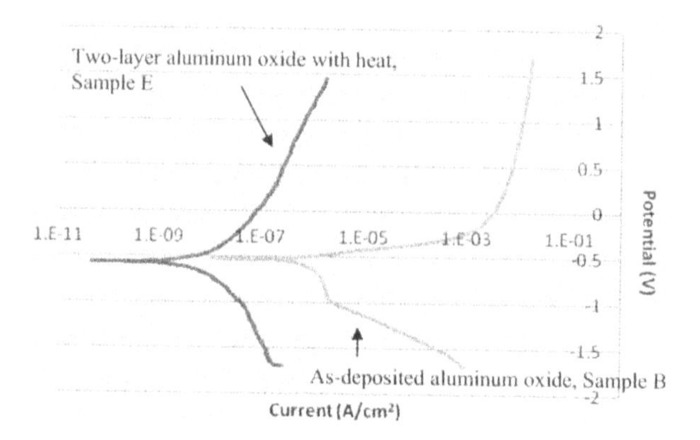

Figure 9 Polarization current vs. potential
Red curve: Two layer aluminum oxide with heat;
Green curve: as-deposited aluminum oxide

To try to further increase the impedance of the heat-treated aluminum oxide coated sample, a UV-curable cerium oxide layer was applied over the aluminum oxide film. The goal was to combine layers of ceramics with a polymer to possibly achieve greater coverage of the substrate as well as improve the performance in anti-corrosion testing.

In Figure 10, the impedance modulus is plotted as a function of frequency for a combined coating which consists of a heat-treated/double-layered aluminum oxide thin film as the bottom layer and UV-curable cerium oxide material as the top layer. With the help of the UV-curable material coating, the impedance modulus of the combined coating is $2x10^6$ Ohm-cm^2.

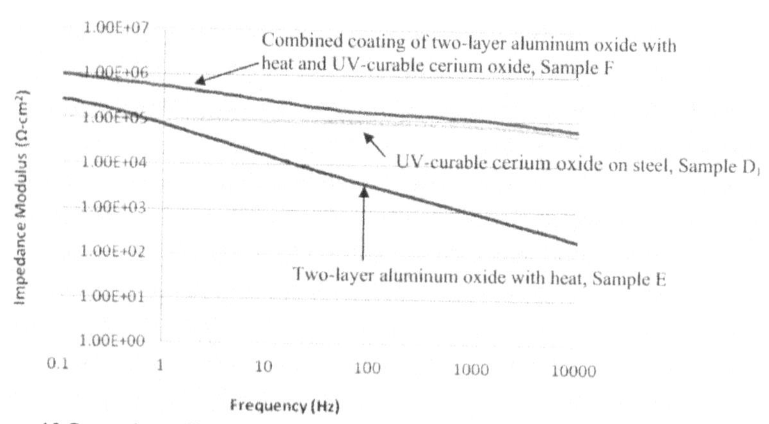

Figure 10 Comparison of impedance modulus for heat treated aluminum oxide film, UV-curable material and the combination.

The UV-curable cerium oxide materials were coated as the top layer. At 0.1 Hz, the impedance modulus reached 10^6 Ohm-cm^2 that was comparable with that of silicon nitride coatings (Figure 11). Presumably, the UV-curable materials on the aluminum oxide film covered some or most of the remaining defect areas.

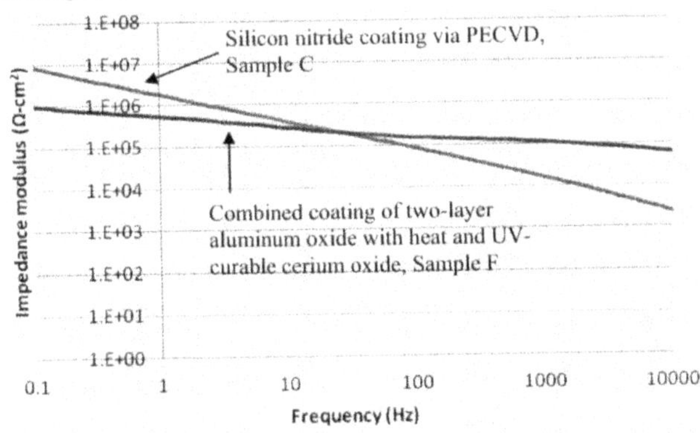

Figure 11 Comparison of impedance modulus of the combined coating and silicon nitride coating.

Among all the thin film materials, silicon nitride thin films can provide the best protection, that is, they can cover carbon steel substrates completely and can resist both general corrosion and galvanic corrosion. In comparison, both aluminum oxide and cerium oxide thin films cannot provide the same or overall protections as the silicon nitride films. Thus, silicon nitride thin film coating is selected as the first layer on the carbon steel. On the top of the first layer, aluminum oxide thin film coating can cover the silicon nitride grain boundaries to further enhance the protections. Thereafter, aluminum oxide and/or cerium oxide embedded UV curable polymer materials can form a hydrophobic top layer. In this paper, we report the new results on the corrosion testing for such three layer configurations.

Zero resistance ammeter test results

The ZRA (zero resistance ammeter) test was used to evaluate the current between magnesium and uncoated steel, aluminum and coated steel. The ZRA test provides an assessment of the insulating potential of the coatings, which is contrasted against aluminum; the preferred means of isolating steel from magnesium in automotive assemblies.

In this set-up, the ZRA current for the control sample (1050 carbon steel plate without any coatings) is found to be 8.83 mA/cm^2 (as tabulated in Table 3). When the carbon steel plate is coated with single-layer aluminum oxide or UV curable coating, the ZRA currents are found to be 8.12 mA/cm^2 and 8.05 mA/cm^2, respectively. (UV curable aluminum oxide, cerium oxide and their mixture are utilized with similar results.) When aluminum oxide (bottom layer) and UV curable aluminum oxide or cerium oxide coating (top layer) are combined in a double layer coating, the current is 6.95 mA/cm^2 . However, when the carbon steel plate is coated with single layer silicon nitride, the ZRA current is found to be 0.79 mA/cm^2.

A triple-layer coating consisting of a bottom layer of silicon nitride, a middle layer of aluminum oxide and a top coat of UV curable cerium oxide, the top layer applied to carbon steel plates resulted in a ZRA current of 0.55 mA/cm^2. This represented the lowest current tested and less than half that of the aluminum baseline, see Table 3, and suggests that the three layer coating could be substituted for aluminum and provide reduced galvanic corrosion.

Table 3 Galvanic Current Measurement Test Results

ID	Sample	Corrosion Current (mA/cm^2)
A	Un-coated Steel (Control)	8.83
B	Aluminum oxide	8.12
C	Silicon nitride	0.79
D_2	UV curable aluminum oxide	8.05
G	Aluminum (Baseline)	1.26
H	Silicon nitride| Aluminum oxide	1.2
I	Silicon nitride | UV curable aluminum oxide	1.1
J	Aluminum oxide |UV curable aluminum oxide	6.95
K	Silicon nitride | Aluminum oxide | UV curable aluminum oxide	0.55

Salt mist spray tests results

The ZRA (zero resistance ammeter) should provide some hints for salt mists spray testing results. Salt spray testing was applied using the test configuration described previously. Aluminum was again used as a baseline and, as shown in Table 4, there is some degree of galvanic corrosion between aluminum and magnesium. Based on the ZRA test results, the three layer coatings were expected to perform better than aluminum with less magnesium mass loss. The salt spray results are reported as both visual inspection (photos) and mass loss percentage on the magnesium plate due to corrosion, both general and galvanic.

In Table 4, photographs of some samples after the testing, and exposed magnesium plate surface via sand-blasting are shown. The photos in the third column show the corrosion right after the 3-day salt mist spray tests. The photos in the fourth column show the corroded area after the corrosion product is removed by sand-blasting. After three days, The Mg plate coupled with carbon steel bolt is severely corroded and a deep "grove" formed around the bolt (Sample 1). Obvious "ring-like" corrosion features are also observed around the position where the aluminum bolt was (Sample 2). For the double layer coated samples (Sample 3 and Sample 4), the galvanic corrosion is less but still observable. However, the bolt of sample 4 shows much more red rust on top of it. There is almost no visible galvanic corrosion feature on the Mg plate coupled with a triple-layer coated carbon steel bolt and the bolt itself shows little corrosion (Sample 5).

Table 4 Visual inspection of sample after 3-day salt mist spray test

ID	Coating	Before Sand-blasting	After Sand-blasting
1	Carbon steel control	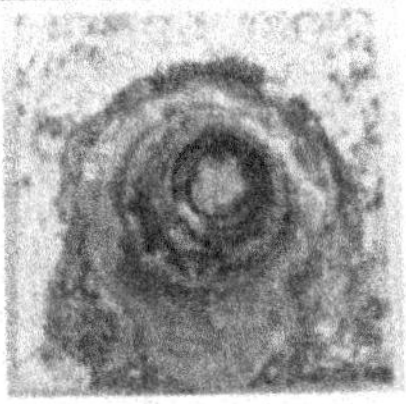	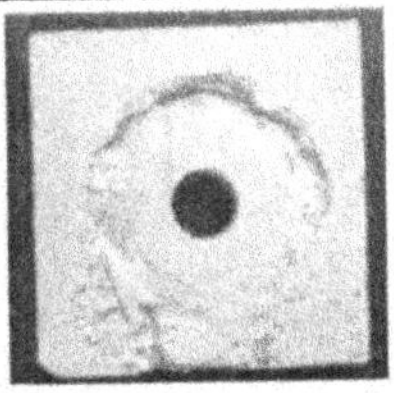
2	Aluminum	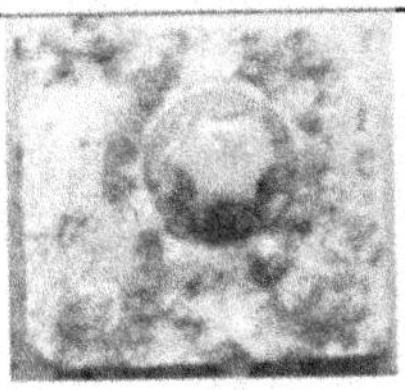	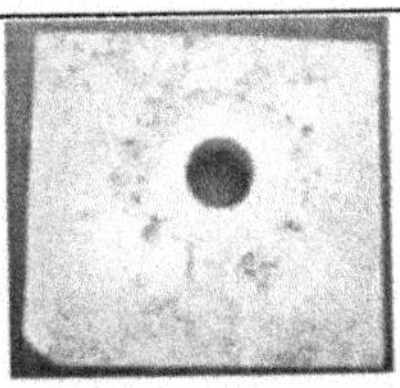
3	2-layer coating: silicon nitride base coat UV-curable aluminum oxide top coat	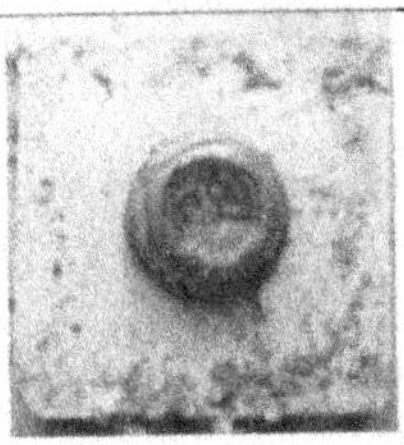	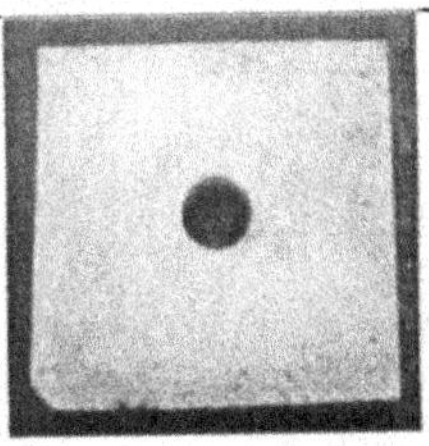
4	2-layer coating: aluminum oxide base coat UV-curable aluminum oxide top coat	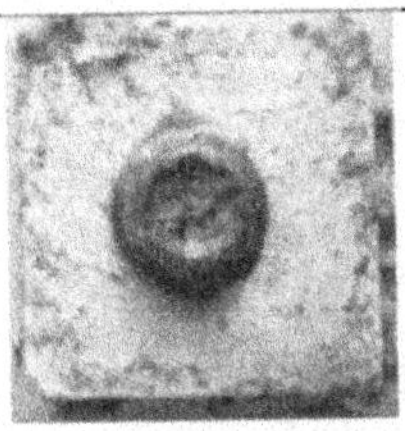	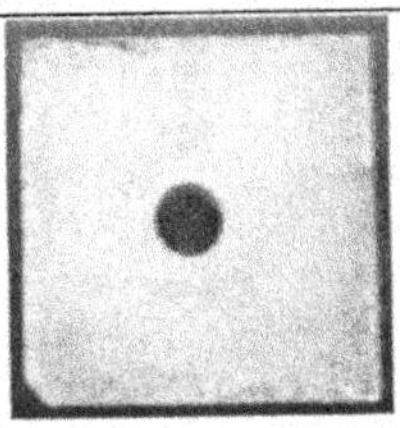
5	3-layer coating: silicon nitride base coat aluminum oxide middle coat UV-curable aluminum oxide top coat		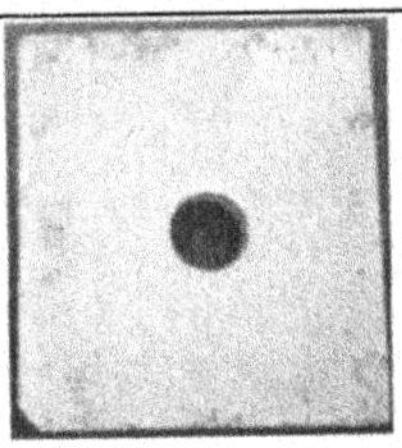

Mass loss

The data of mass loss of the Mg plates confirmed the visual examination. The mass loss for Mg plate coupled with the triple-layer coated carbon steel bolt is much less than that coupled with the control and aluminum bolt. See Table 5.

Table 5 Mass loss % after 3-day salt mist spray test

ID	Sample	Mass Loss (%)
1	Carbon steel bolt	3.32
2	Aluminum bolt	0.65
3	2-layer coating: silicon nitride base coat UV-curable aluminum oxide top coat	0.24
4	2-Layer Coating: aluminum oxide base coat UV-curable aluminum oxide top coat	0.48
5	3-layer coating: silicon nitride base coat aluminum oxide middle coat UV-curable aluminum oxide top coat	0.08

It is shown that the triple layer coated steel bolts perform better than the aluminum bolts in salt mist corrosion testing. This general trend is consistent with the earlier observation.[9]

CONCLUSION AND DISCUSSION

The ZRA (zero resistance ammeter) measurements indicate that the galvanic corrosion current reduces significantly from 8.83 mA/cm^2 for uncoated carbon steel plate to 0.55 mA/cm^2 for 3-layer coated carbon steel plate. The reduced current is even smaller than that of bare Al-Mg couple. The salt mist spray testing demonstrates that the 3-layer coated steel bolt performs better than the aluminum bolt as illustrated in Tables 4-5. Thus, 3-layer coating reduces the galvanic current between the coated steel and magnesium in a 5% salt water solution. This suggests that the coated steel fasteners may be a viable replacement to more expensive aluminum isolation for joining of magnesium. Currently, we are conducting a follow-up research project to see if 3-layer coatings can be applied sequentially in one deposition facility at Southwest Research Institute. In particular, the deposition technique is similar to that of PECVD. The first two coatings will also be similar to that reported here. However, the third layer will be a silicon oxide doped DLC (diamond like carbon). Preliminary studies show that the silicon oxide doped DLC layer does display some characteristics of that offered by the UV curable layer.

ACKNOWLEDGEMENT

Help provided by J. Dalbey and T. Rein is appreciated.

DISCLAIMER

This material is based upon work supported by the Department of Energy National Energy Technology Laboratory under Award Numbers DE-FC05-02OR22910, DE-FC26-02OR22910 and DE-EE0003583. This research is prepared as an account of work sponsored by an agency of the United States Government. Neither the United States Government nor any agency thereof, nor any of their employees, makes any warranty, express or implied, or assumes any legal liability or responsibility for the accuracy, completeness, or usefulness of any information, apparatus, product, or process disclosed, or represents that its use would not infringe privately owned rights. Reference herein to any specific commercial product, process, or service by trade name, trademark, manufacturer, or otherwise does not necessarily constitute or imply its endorsement, recommendation, or favoring by the United States Government or any agency thereof. The views and opinions of authors expressed herein do not necessarily state or reflect those of the United States Government or any agency thereof. The partial financial support from USCAR program is acknowledged.

BIBLIOGRAPHY

1. USAMP (United States Automotive Materials Partnership) (2006) Magnesium Vision 2020.
2. G. Song, B. Johannesson, S. Hapugoda, and D. StJohn, "Galvanic Corrosion of Magnesium Alloy AZ91D in Contact with an Aluminum Alloy, Steel and Zinc," Corrosion Science, Vol. 46, 2004, p.p 955–977.
3. J. X. Jia, G. Song, and A. Atrens, "Influence of Geometry on Galvanic Corrosion of AZ91D Coupled to Steel," Corrosion Science, Vol. 48, 2006, p.p. 2133–2153.
4. Y. Liu, C. Qu, R. E. Miller, D. D. Edwards, J. H. Fan, P. Li, E. Pierce, A. Geleil, G. Wynick, and X. W. Wang, "Comparison of Oxide and Nitride Thin Films – Electrochemical Impedance Measurements and Materials Properties," Ceramic Transaction, Vol. 214, 2010, p.p. 131-146.
5. C. Qu, P. Li, J. Fan, D. Edwards, W. Schulze, G. Wynick, R. E. Miller, L. Lin, Q. Fang, K. Bryson, H. Rubin, and X. W. Wang, "Aluminum Oxide and Silicon Nitride Thin Films as Anti-Corrosion Layers," Ceramic Engineering & Science Proceedings, Vol. 31, Issue 3, 2010, p.p. 123-134.
6. L. Lin, C. Qu, R. Kasica, Q. Fang, R. E. Miller, E. Pierce, E. McCarty, J. H. Fan, D. D. Edwards, G. Wynick, and X. W. Wang, "Multilayer Coatings for Anti-Corrosion Applications," Ceramic Engineering & Science Proceedings, Vol. 32, Issue 3, 2011, p.p. 59-70.
7. Y. Liu, "Silicon Nitride and Other Oxide Thin Film on Carbon Steel Substrates - Anti-Corrosion Barrier Layer Application"; M.S. Thesis. Alfred University, Alfred, NY, 2008.
8. C. Qu, "Aluminum Oxide Thin Films as Anti-Corrosion Layers"; M. S. Thesis. Alfred University, Alfred, NY, 2009.
9. L. Lin, "Evaluations of Anti-corrosion Behavior for Multilayer Thin Film Coatings"; M. S. Thesis. Alfred University, Alfred, NY, 2010.

ACOUSTIC FIELD ASSISTED DRYING OF ELECTROPHORETICALLY DEPOSITED COLLOIDAL COATINGS

Changzheng Ji and Houzheng Wu

ABSTRACT

Ceramic green coatings with a thickness from tens microns to a few millimetres can be manufactured through electrophoretic deposition (EPD) on various substrates with clear merits in economic and technical sides. However, drying of as-deposited colloidal coatings is often problematic in restraining possible cracking likely induced by the in-plane tensile introduced by shrinking during the evaporation of solvent from the drying coatings. It was witnessed that for a given range of particle size and substrate, there existed a critical thickness that above it, cracking started; underneath it, no cracking appeared during drying. With the assistance of acoustic field, generated by an ultrasonic generator, the aforesaid cracking was significantly mitigated and the critical thickness was notably raised. In this paper, the effect of particle sizes on the critical thickness was quantified under chosen depositing and drying conditions with and without acoustic field. Possible mechanisms on the development of drying cracking and the role of acoustic field in mitigating this phenomenon were discussed.

1. INTRODUCTION

Drying ceramic coatings produced from colloidal techniques is a critical stage for a successful coating and has been interested by researchers. When a colloidal coating is deposited on a non-porous rigid substrate followed by drying, the evaporation of solvent in the drying coating leads to a reduction of volume and a concentration of solid particles to generate a compact or network of particles/agglomerates. For a drying coating on a rigid substrate, any in-plane shrinkage is likely constrained by the cohesion between the drying coating and substrate, and lead to a tensile stress inside the drying coating. Such a tension can be accommodated by matter transportation; whenever this transportation is not enough or no longer available to accommodate the required displacement, which becomes the situation near the end of drying, a tensile stress would be built up inside the drying coating. When the stressing level exceeds the cohesive strength of the compact, cracking will happen inside the drying coating.

The development of stresses in drying coatings on rigid substrates has been extensively discussed [1-10]. In-plane tensile stresses, which are induced by the constraint from the rigid substrate, start to build up due to the restriction on movement of particles in the vicinity of the substrate. The stresses can be released by cracking, if they exceeded the strength of the powder compact. It has been reported [7, 8, 11] that for drying coatings on a rigid substrate, cracks only occurred in coatings whose thickness are higher than a certain value, known as the critical cracking thickness (CCT). Lange [11] has theoretically explained the existence of a CCT in a coating:

$$d_c = \frac{2GE}{K\sigma^2(1-\nu)} \tag{1}$$

where d_c is the CCT, G the energy needed to form two surfaces of a crack per unit area (which is known as the critical strain-energy release rate in fracture mechanics), K is a dimensionless parameter, defined by the type of cracks [14], σ the tensile stress in the coating, E and ν the Young's modulus and Poisson's ratio of the coating respectively. This model implies that both tensile stresses built up by the drying and fracture toughness of the coating preform determine the critical thickness of cracking. To reduce the tensile stress, Bauer and Cima [2] experimentally proved that an improved flux of liquid in the drying body resulted in a uniform drying without cracking. Lan and Xiao [13] reported a case where the tensile stress could be significantly reduced when large particles were mixed with fine ones.

Descamps et al [14] demonstrated that the reduction of drying rate could lead to a larger CCT and vice versa; they believed that slow drying rate had tensile stress relaxed through mass transportation within the coating.

The effectiveness of an acoustic field on drying of materials has been recognised since 1936 [15]. Researchers [16, 17] have suggested that the acoustic field could promote the drying rate by improving the flow of solvent in capillaries of the drying body. Theoretical researches [18-20] also supported that the pressures applied by an acoustic radiation on existing particles can be predicted by mathematical models. All those demonstrated possible interactions of acoustic field on colloidal slurries, which could benefit the drying of colloidal coatings. We therefore envisage that such assistance may mitigate the cracking during drying and improve the CCT of drying coatings.

In this study, we prepare coatings with various thicknesses using electrophoretic deposition (EPD). The focus is to investigate the potential benefit that acoustic field can provide on the drying process of EPD coatings through quantifying CCT and comparing the development of cracks during drying.

2. EXPERIMENTAL

2.1. Materials

Silicon powder (99%, Sigma-Aldrich) was used as starting material for coating preparation. Acetone (99.5%, Fisher Scientific) was chosen as dispersing solvent of suspensions for the depositions. Tetramethyl-ammonium (TMA) hydroxide (25wt% in water, Sigma-Aldrich) was added into the suspension to provide negative charge on the surface of silicon particles dispersed in acetone. Stainless steel was used as the substrate of coating and machined into disc shape with a diameter of 25mm and a thickness of 5 mm. The substrate surfaces were finished with SiC sand paper (P1200, Struers) to achieve identical surface topography before depositing coatings.

2.2. Coating preparation

20.0 gram of as-received silicon was milled in an attritor (01-HD, Union Process) in acetone (200 ml) with 0.5 ml TMA solution added. The attrition milling was operated at rotating rate of 300 rpm; both the arm and the liner of the container were made of Y-TZP, and the milling media was tungsten carbide. Suspensions including different size distributions of Si particles were achieved by controlling the milling time. Three sets of suspensions were prepared by setting milling duration as 1, 2 and 3 hours. The as-milled suspensions were further diluted by directly adding acetone, two times volume of the original suspension, to the suspension followed by mechanically stirring for 30 minutes and ultrasonically treating for 30 minutes. The diluted suspensions, with an estimated solid concentration of 0.3 g/ml, were ready for the EPD process.

Electrophoretic deposition (EV265, Consort) was conducted on the surface of stainless steel substrates in the as-prepared Si suspension. In order for comparing the drying conditions of as-deposited colloidal coatings, two substrates connected with copper conduction wire were set as the targeting electrode. Such a setup is expected to achieve a coating on each substrate with exactly same thickness and structure of constituents. The counter electrode was a stainless steel disc with a diameter of 60 mm and a thickness of 1 mm. The distance between the electrodes was 30 mm. A schematic drawing of the EPD setup is shown in Fig. 1.

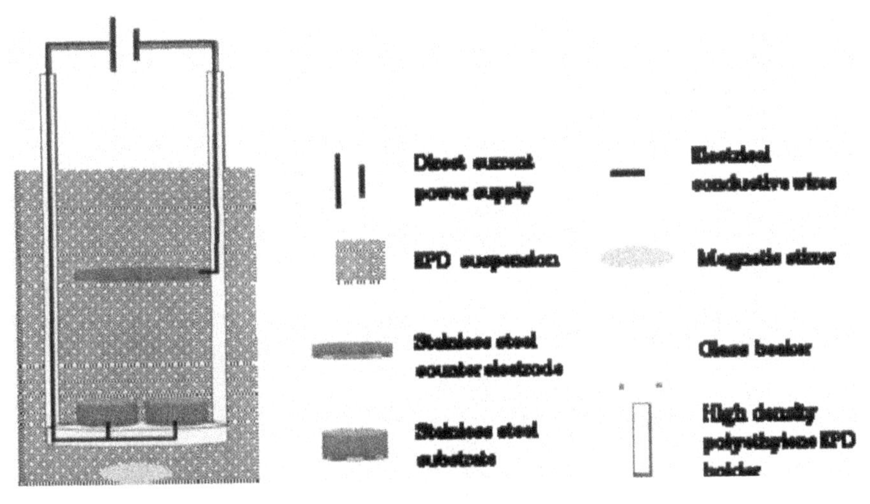

Figure 1. Schematic drawing of EPD setup.

The potential for all EPD exercises was set as 120 V. The thickness of the deposition was defined by controlling the deposition time ranging from 20 sec to 10 min.

2.3. Drying coatings

Each freshly deposited colloidal coating, together with the substrate, was dried in a standard glass beaker with a volume of 100 ml (50 mm in diameter × 70 mm in height, Fisherbrand) where a 5.0 ml of acetone was loaded. A part of the substrate was merged in acetone to keep the drying environment relatively consistent and stable, whereas the coating on top of the substrate was kept away from liquid acetone with a distance of > 2 mm. The beakers were left in open during drying, while they were merged in a water bath at room-temperature, typically ranging between 18 °C to 22 °C. For each set of direct comparison, however, the temperature difference of the water bath had no more than 2 °C. For acoustic-field assisted drying, the field was implemented by loading the beaker into an ultrasonic bath produced by a generator with a frequency set as 25 kHz and output power 250 W (UCE Ultrasonic). At least half of the beaker was emerged in the ultrasonic bath.

2.4. Determination of critical cracking thickness (CCT)

To identify CCT, preparation of coatings with variable thickness is briefed as follows. 20 seconds was set as a unit for changing the time of deposition. Based on some trials, for P1, P2 and P3 suspension, the starting deposition time for the first coating was set as 4, 2 and 1 min. respectively. The as-deposited colloidal coating was dried following the procedure given in section 2.3. If the first coating had any drying cracks, then the deposition time for the following coating would be reduced by 20 sec., otherwise 20 sec. would be added. This process continued until ten coatings deposited were crack free. Then, the average thickness of the ten coatings was set as the CCT of a suspension under given drying conditions. In this study, only visible cracks by naked eyes were used to determine CCTs, whereas micro-cracks were not taken into account on this occasion.

2.5. Characterisation

The size distribution of Si suspension prepared by the attritor was measured using a laser diffraction technique (Mastersizer 2000, Malvern). The particle shape and the agglomeration were observed using a field-emission gun scanning electron microscopy (FEG-SEM) (Carl Zeiss 1530 VP, Germany). The thickness of the coating was measured after drying using an optical microscope (VHX-600, Keyence) with a 3-D imaging system. The crack patterns were recorded using a digital camera. For each set of experiments, at least three samples were repeated, and the results were averaged with standard deviations available for all numerical measurements.

3. RESULTS

3.1. Powder characterisation

Fig. 2 shows the particle size distributions (PSD) of silicon in suspensions before and after milling for different periods. The centroid positions of the primary peaks (the ones have typical normal distribution) of Si particle size distribution present a clear trend of decreasing from 4.4 μm, to 3.7, 2.0 and 1.4 μm as milling time increased from 0, to 1, 2 and 3 hrs. The distribution of agglomeration had subjected to a substantial change. In the as-received powder, only a small amount of agglomeration and/or large particles appeared in a size regime around several hundreds microns; after milling for 1 hr, apart from the original agglomerates/large particles, a new range of agglomerates/large particles were formed in a size regime of around tens microns. By extending the milling to 2 and 3 hrs, the amount of agglomerates/large particles increased rapidly and the size regime ranging from tens to hundreds microns.

SEM images in fig. 3 directly illustrate the size, shape and agglomerating of Si particles before and after attrition milling. Apart from echoing the reduction trend of Si particles in the fine regime of the size spectrum, the important information is that the agglomerates and large crystallites inside the regime of large size can be differentiated. In the as-received powder, large crystallites were in dominant with no agglomeration; after 3 hrs milling, it was agglomerates in dominant, and very few large Si crystallites were noticed with a comparable size of large agglomerates. After milling for 1 or 2 hrs, both agglomerates and large Si crystallites co-existed with comparable sizes.

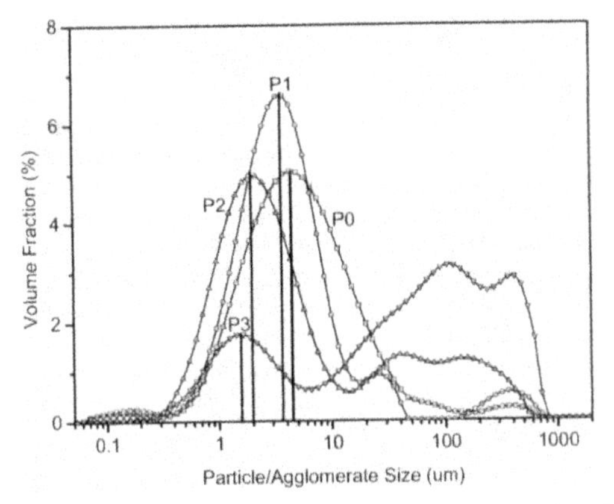

Figure 2. Particle/agglomerate size distribution of silicon in suspension. As-purchased powder P0 and this after milling for 1 hr (P1), 2 hrs (P2) and 3 hrs (P3).

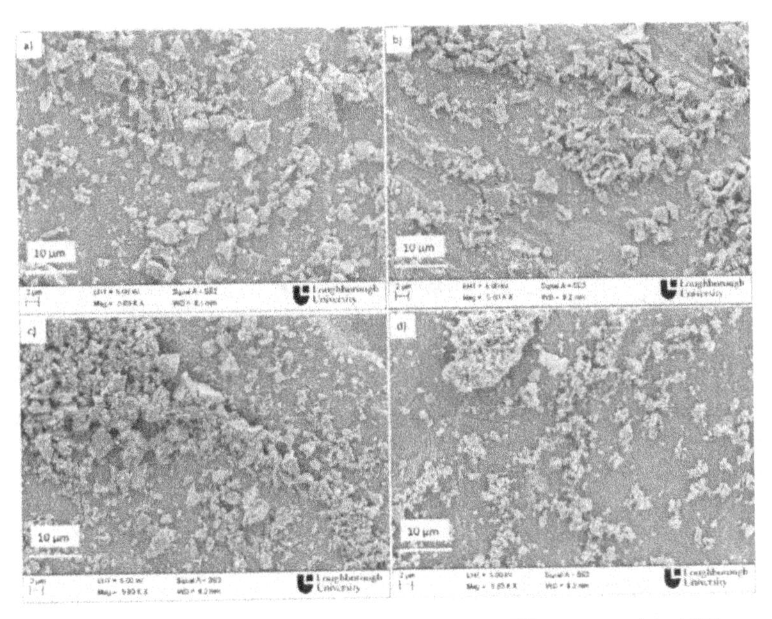

Figure 3. SEM images of silicon powders before and after milling: a) as-purchased (P0), b) after milling for 1 hr (P1), c) after milling for 2 hrs (P2), d) after milling for 3 hrs (P3).

3.2. EPD of colloidal coatings

For the purpose of producing colloidal coatings with high reproducibility and uniform and controllable thickness, the EPD conditions were fixed, including the concentration of particles and additives in the suspension, the applied voltage, the deposition area of substrates and the distance between the substrate and the counter electrode. The only variable during EPD for controlling the coating thickness was the deposition time. Fig. 4 shows the increasing trend of thickness of the coating with the depositing time for all three suspensions. Note, the thicknesses indicated in the plot were measured from green coatings after drying fully under ambient conditions. It is clearly indicated that a thicker coating was achieved when the deposition was conducted in a suspension containing larger particles (e.g. P1 in fig. 4) under current processing conditions. The arrows in fig. 4 indicate a thickness where cracking started to appear during drying; the value of the thickness is close to the critical cracking thickness of coatings that was deposited from each suspension. For P1, P2 and P3 suspension, the value of CCT was no more than ~1100 μm, ~480 μm, and ~140 μm respectively.

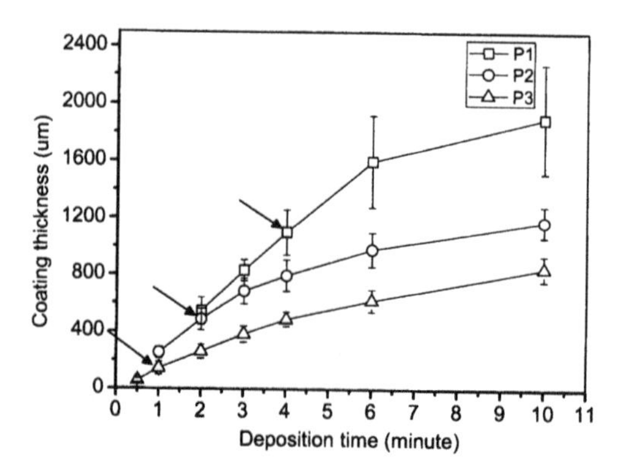

Figure 4. EPD coating thickness as a function of deposition time.

3.3. CCT of EPD coatings

Since the measurement of CCT was based on a visual observation of the appearance of cracks during drying, the reported values have statistic meaning, instead of a clear cutting-off value, as described in section 2.4.

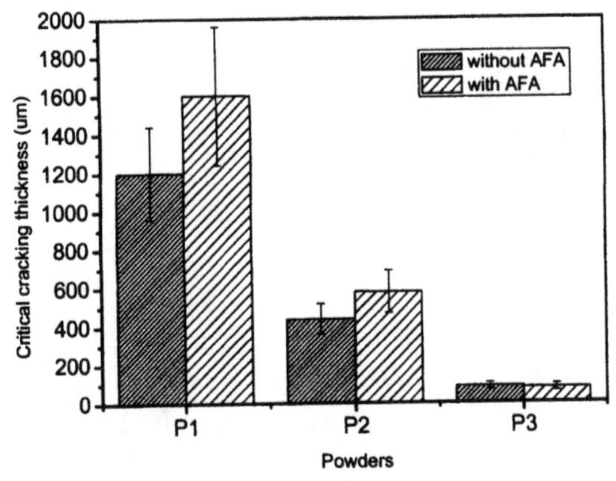

Figure 5. Averaged critical cracking thickness of coatings produced in difference suspensions and dried with and without the assistance of acoustic field.

The averaged measurements of CCT were shown in Fig. 5 for coatings deposited in different suspensions. The coatings dried with the assistance of acoustic field had higher CCT values than those dried without such field. The difference became more obvious when relatively larger particles were included in the suspension. To be more specific, for coatings prepared from P1 suspension, the CCT value increased from 1200 μm to 1600 μm when acoustic field was applied. For those from P2 suspension, the increment of CCT was about 140 μm when drying was assisted by acoustic field; for those from P3 suspension, little impact of acoustic field was observed on the CCT value.

3.4. Cracks developed in the green coatings

In the interest of addressing the impact of AFA on the development of drying cracks, pairs of coatings were prepared under identical processes, as described in session 2.2, with a fixed EPD deposition time of 10 min. The corresponding thicknesses can be obtained from fig. 4. For each particular pair of coatings, their thicknesses were measured to share the same value within ±2% measurement error. Images in fig. 6 show the green coatings after drying with and without support of acoustic field. For coatings from P1 suspension, their thicknesses were just inside the regime of CCT. After drying, no cracking appeared with AFA applied, while a large crack appeared in the green coating without AFA applied (fig. 6 a) and d)). Coatings from P2 suspension, the thickness was expected to exceed the CCT. Cracks appeared in coatings regardless AFA applied or not. For coatings made from P3 suspension, the thickness was well above the CCT, and extensive cracking appeared under both drying conditions. It is noted that in green coating with AFA applied, the crack network consisted of short, narrow and curved cracks (fig. 6 c)), and in that without AFA applied, cracks in the network were long, wide and straightened (fig. 6 f)).

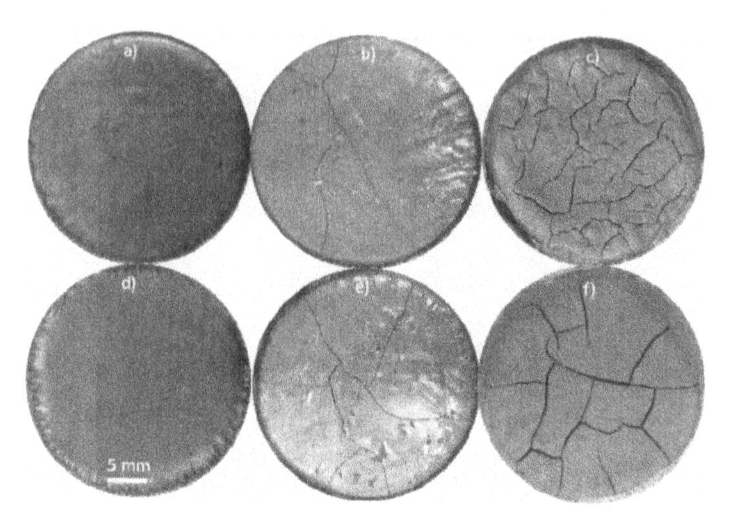

Figure 6. Images of coatings showing crack patterns after drying with, a), b) and c) and without d), e) and f) AFA. Coatings in a) and d) were prepared from P1 suspension, those in b) and e) from P2 suspension, and those in c) and f) from P3 suspension.

4. DISCUSSIONS

Our results have demonstrated that the application of an acoustic filed during drying can improve the critical thickness for cracking of colloidal coatings deposited on a rigid substrate. The potential benefits brought by the acoustic field could be understood, as per the improvement of particle packing and solvent flowing in the drying coating.

(a) Improvement on particle packing by acoustic field

It has been mathematically described [19] that with application of an acoustic field, particles in the field experience time-averaged forces from the interactions between particles and the acoustic wave field. Fig. 7 schematically shows the forces applied on a particle by the acoustic waves.

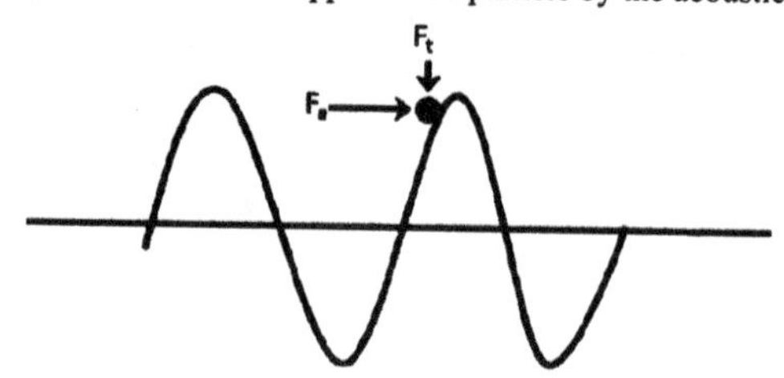

Figure 7. Schematic drawing of forces applied on a particle by an acoustic wave.

The axial radiation force, F_a, shakes the particle on the axial direction of the wave, and the transverse force, F_t, provides vertical pressures on the particle. F_a was derived by Yosioka and Kawasima [20] as below:

$$F_a = V_0 J_a \frac{2\pi}{\lambda} C \sin(4\frac{\pi}{\lambda}x)k \tag{2}$$

where V_0 is the volume of the particle, J_a the acoustic energy density, λ the sound wavelength, x the axial distance from a pressure node, k the unit vector in the axial direction, and C is an acoustic contrast factor, which reflects the difference in compressibility and density between the particle and the solvent.

From equation (2), for a given acoustic field, the magnitude of the force F_a is proportional to the size of the particle, and its direction varies with the wave propagation. Effectively, the acoustic field applies a vibration under certain frequency on the particles in the drying slurry. We believe the vibration increases the mobility of solid particles that promotes the matter transportation during drying to accommodate any tension generated at varies scales. The outcomes from drying coatings made from P1 and P2 suspensions have demonstrated the benefit that acoustic field can provide. However, we did notice that the effectiveness faded away with the reduction of particle size. Actually, coatings made from P3 suspension that had the finest particle among the three samples did not show any benefit by applying an acoustic field during drying. Only small force can be applied on smaller particle, according to eq.(2), which may be part of the reasons that only limited mobility could be boosted by the acoustic field for drying coating consisting of very fine particles. However, we believe other factors may have impact on the mobility boosting by acoustic field, such as its frequency and the existence of agglomeration, which needs further understanding.

(b) Improving solvent transport in the drying coating

Biot [21] predicted a model to claim that the liquid flow in small channels can be significantly improved from viscous flow to plug flow, if there is an elastic wave, which has a frequency, f, higher than a certain value, applied on the fluid. The frequency value is given by:

$$f \geq \frac{\pi\eta}{4\rho d^2} \tag{3}$$

where η and ρ are, respectively, the viscosity and the density of the flowing liquid, d the diameter of the channel, which is proportional to the particle size in the drying body [22]. For the present research, the solvent of acetone has a viscosity of 3.31×10^{-4} Pa·s and density of 0.79 g·cm^{-3} at 20 °C [23]. The particle/agglomerate size ranges from 0.3 μm up to 1000 μm (fig. 2), which could give channels with an equivalent diameter of approximately from 0.1 μm to 400 μm. Putting these values into eq. (3), the required minimum frequency of elastic wave should be in the range from 20 kHz to 330 GHz. This estimation implies that flowing of solvent in very large channels could be improved under the current frequency of acoustic wave. Clearly, large particles can provide large channels where the solvent flow inside can be improved by the applied field. Our tests did indicate that coating containing coarser particles experienced much better response on the acoustic field in prevent cracking. Unfortunately, the flowing of solvent during drying was not monitored inside the coating to firmly validate the impact of elastic wave predicted by eq. (3).

5. CONCLUSIONS

The benefits of acoustic field on drying colloidal coatings were experimentally demonstrated. The critical cracking thickness of coatings was significantly improved when the coatings were prepared by EPD technique in a ceramic suspension containing relatively large particles. As the particle sizes in the suspension became finer, the improvement in critical cracking thickness faded away. It is believed that the application of acoustic field during drying can reinforce the packing of particles and the flowing of solvent when the parameters in the coating and frequency of acoustic field are appropriately matched.

REFERENCES

[1] E. Santanach Carreras, F. Chabert, D. E. Dunstan and G. V. Franks, "Avoiding 'mud' cracks during drying of thin films from aqueous colloidal suspensions", J. Colloid Interface Sci., 313, 160-168, 2007

[2] C. Bauer, M. Cima, A. Dellert and A. Roosen, "Stress development during drying of aqueous zirconia based tape casting slurries measured by transparent substrate deflection method", J. Am. Ceram. Soc., 92 (6), 1178-85, 2009

[3] W. Lan and P. Xiao, "Constrained drying of aqueous yttria-stabilized zirconia slurry on a substrate I: Drying mechanism", J. Am. Ceram. Soc., 89 (5), 1518-22, 2006

[4] M. S. Tirumkudulu and W. B. Russel, "Cracking in drying latex films", Langmuir, 21, 4938-48, 2005

[5] J. A. Lewis, K. A. Blackman, A. L. Ogden, J. A. Payne and L. F. Francies, " Rheological properties and stress development during drying of tape-cast ceramic layers", J. Am. Ceram. Soc., 79 (12), 3225-34, 1996

[6] C. J. Martinez and J. A. Lewis, "Rheological, structural and stress evolution of aqueous Al_2O_3: Latex tape-cast layers", J. Am. Ceram. Soc., 85 (10), 2409-16, 2002

[7] R. C. Chiu, T. J. Garino, and M. J. Cima, "Drying of granular ceramic films I: Effect of processing variables on cracking behaviour", J. Am. Ceram. Soc., 76 (9), 2257-64, 1993

[8] R. C. Chiu and M. J. Cima, "Drying of granular ceramic films II: Drying stress and saturation uniformity", J. Am. Ceram. Soc., 76 (11), 2769-77, 1993

[9] W. Lan, X. Wang and P. Xiao, "Agglomeration effect on drying of yttria-stabilised-zirconia slurry on a metal substrate", J. Eur. Ceram. Soc., 26 (16), 3599-3606, 2006

[10] W. Lan and P. Xiao, "Drying stress of yttria-stabilized-zirconia slurry on a metal substrate", J. Eur. Ceram. Soc., 27, 3117-25, 2007

[11] F. F. Lange, "Chemical solution routes to single-crystal thin films", Science, 273, 903-09, 1996

[12] J. W. Hutchinson and Z. Suo, Advances in Applied Mechanics, 63-191, Academic Press, New York, 1991

[13] W. Lan and P. Xiao, "Constrained drying of an aqueous yttria-stabilized zirconia slurry on a substrate II: Binary particle slurry", J. Am. Ceram. Soc., 90 (9), 2771-78, 2007

[14] M. Descamps, M. Mascart and B. Thierry, "How to control cracking of tape-cast sheets", Am. Ceram. Soc. Bull., 74 (3), 89-92, 1995

[15] F. J. Burger and K. Söllner, "The action of ultrasonic waves in suspensions", Trans. Faraday Soc., 32, 1598-1603, 1936

[16] P. Greguss, "The mechanism and possible applications of drying by ultrasonic irradiation", Ultrasonics, 1 (2), 83-6, 1963

[17] H. V. Fairbanks, "Ultrasonically assisted drying of fine particles", Ultrasonics, 12 (6), 260-2, 1974
[18] Louis V. King, "On the acoustic radiation pressure on spheres", Proc. R. Soc., 147, 212-240, 1934
[19] N. Aboobaker, D. Blackmore and J. Meegoda, "Mathematical modelling of the movement of suspended particles subjected to acoustic and flow fields", Appl. Math. Model., 29, 515-32, 2005
[20] K. Yosioka and Y. Kawasima, "Acoustic radiation pressure on a compressible sphere", Acoustics, 5, 167-73, 1955
[21] M. A. Biot, "Theory of propagation of elastic waves in a fluid saturated porous solid I. Low-Frequency Range", J. Acoust. Soc. Am., 28 (2), 168-78, 1956
[22] J. Reed, Introduction to Principles of Ceramic Processing, John Wiley & Sons Inc., NY, 1995
[23] David R. Lide, CRC Handbook of Chemistry and Physics, 73rd edition, CRC Press Inc., 1992–1993

Thermal Protection Systems

TESTING OF CANDIDATE RIGID HEAT SHIELD MATERIALS AT LHMEL FOR THE ENTRY, DESCENT, AND LANDING TECHNOLOGY DEVELOPMENT PROJECT

Steven Sepka[1]
ERC Incorporated, Huntsville, AL, 35805

Matthew Gasch,[2] Robin A. Beck,[3] and Susan White[4]
NASA Ames Research Center, Moffett Field, CA, 94035

ABSTRACT
The material testing results described in this paper were part of a material development program of vendor-supplied, proposed heat shield materials. The goal of this program was to develop low density, rigid material systems with an appreciable weight savings over phenolic-impregnated carbon ablator (PICA) while improving material response performance. New technologies, such as PICA-like materials in honeycomb or materials with variable density through-the-thickness were tested. The material testing took place at the Wright-Patterson Air Force Base Laser Hardened Materials Laboratory (LHMEL) using a 10.6 micron CO_2 laser operating with the test articles immersed in a nitrogen-gas environment at 1 atmosphere pressure. Test measurements included thermocouple readings of in-depth temperatures, pyrometer readings of surface temperatures, weight scale readings of mass loss, and sectioned-sample readings of char depth. Two laser exposures were applied. The first exposure was at an irradiance of 450 W/cm^2 for 50 or 60 seconds to simulate an aerocapture maneuver. The second laser exposure was at an irradiance of 115 W/cm^2 for 100 seconds to simulate a planetary entry. Results from Rounds 1 and 2 of these screening tests are summarized.

I. INTRODUCTION

A. Background

Most ablative heat shields are too large to be constructed from a single piece of material. For a multi-piece ablative heats shield, the standard approach to avoid windward-facing gaps is to bond together the various pieces. NASA used this gap bonding method for the Pioneer-Venus and Galileo entry probes, and more recently for Mars Science Laboratory[1] (MSL) that will reach Mars in August 2012. Space-X also used bonding for the Dragon capsule that successfully entered from Earth orbit in December 2010. Both MSL and Dragon use relatively large pieces of phenolic-impregnated carbon ablator (PICA[2]) bonded together with room temperature vulcanizing silicone rubber (RTV-560). An alternative, labor-intensive approach is to cover the forebody structure with a filled honeycomb. This approach was used with Avcoat ablator for Apollo[3] and SLA-561V ablator for all NASA Mars missions prior to MSL.[4] Both bonded and honeycomb heat shields may be described as monolithic.

---

[1] Senior Research Scientist, NASA-Ames Research Center, Thermal Protection Materials Branch (TSM) MS-N234-1, Moffett Field, CA, 94035
[2] Aerospace Engineer, Thermal Protection Materials Branch (TSM), MS-N234-1, Moffett Field CA, 94035.
[3] Aerospace Engineer, Entry Sytems and Vehicle Development Branch (TSS), MS-N229-1, Moffett Field CA, 94035.
[4] Research Engineer, Thermal Protection Materials Branch, Mail Stop 234-1, Moffett Field CA, 94035.

There are some notable exceptions to the two methods of bonding and honeycomb. For example, the Stardust[5] entry probe was sufficiently small such that the forebody heatshield could be manufactured as a single piece of shape-cast PICA. The small DS-2 probes were machined from near-net-shape tiles, and then impregnated with silicone. The Genesis probe used an external carbon shell with large structural attachments that penetrated through a porous carbon insulator.

For bonded and honeycomb heat shields, it is preferential to have the honeycomb or gap filler recede at an equal or only slightly faster rate than the acreage material(s). For aeroheating environments where the ablator recedes faster than the bond/honeycomb, protrusions can develop which may create flow disturbances[6] and heating augmentation. This heating augmentation should be taken into account for sizing the material thickness.

For example, aerothermal environments severe enough for ablation can be found in direct planetary entry with entry high velocity and ballistic coefficients, or in an aerocapture maneuver to decelerate a spacecraft into orbit by using a planet's upper-atmosphere for drag. However, for the later stages of direct entry or for planetary entry after aerocapture, the aerothermal environment is usually mild enough to warrant the use of an insulator. By tailoring the density of TPS materials to have sufficient material for each phase of entry, significant mass savings[7] can be achieved. Recently, a series of arc-jet experiments[8] at NASA Ames were conducted to study this effect. These tests had an ablator, PICA, stacked on top of an insulator, LI-900. Results from this study validated the "dual layer" concept.

An ablator can be considered to consist of three primary parts: a reinforcing material such as metallic honeycomb or ceramic matrix, fillers such as microballoons or cork, and a polymeric binder such as phenolic resin to act as an adhesive for all of the components. [9] The current state-of-the-art for low-density ablators is PICA. In the past, low-density ablators such as this have been used for Earth and Martian planetary entry. One limitation of PICA is that it cannot be fabricated into very large pieces; therefore, a large heat shield may require abutment or bonding. Recent efforts, however, have been to place PICA into a metallic honeycomb.[10] Additionally, although PICA is currently available as only a single density material, secondary impregnation techniques have been developed to densify the outer surface to a depth of approximately 2 cm.

The results presented here were part of a material screening testing program of vendor-supplied, proposed heat shield materials. The goal was to develop materials with better ablative performance than PICA with appreciable weight savings. In particular, new designs using proprietary materials, varying density (i.e. functionally graded) through the material, and the use of honeycomb-based systems will be evaluated for thermal performance.

The usual approach in heat shield design is to determine the amount of heat shield material required to keep a bond line temperature below a desired value when the heat shield is exposed to an aerothermal heat pulse. For the current screening work, the problem was the inverse, i.e., if material thickness was chosen to keep areal density constant, how would the bond line temperature differ between materials. Other properties of interest included char depth (or percentage virgin material remaining), mass loss, and the general response of the material based on visual inspection. This work was part of NASA's Entry Descent and Landing (EDL), Technology Development Project (TDP), Thermal Protection Systems (TPS) Element, Rigid Materials Development task.

B. Test Facility

All tests were conducted at the Wright-Patterson Air Force Base Laser Hardened Materials Evaluation Laboratory[11] (LHMEL). The tests utilized the 15-kW LHMEL I laser, which was a continuous-wave, electric discharge, coaxial, CO_2, gas laser. The materials tested were opaque to the laser's 10.6-micron output wavelength. The laser produced a calibrated, spatially-uniform "flat-top" heat flux profile beam. All tests reported here had a 4.65 cm (1.83 inch) beam diameter at the sample surface. Because the laser radiation was absorbed at the surface of the material, as opposed to in-depth, these tests were well-suited as screening tests.

The LHMEL facility was used to impart a known thermal load onto the material specimens via radiant heating. Because the environmental test chamber was made inert by a continual addition (purge) of nitrogen gas, comparisons of thermal protection capability, without the additional complication of surface recession, were made. Laser ablation, whereby the surface material liquefies and vaporizes due to the energy of the beam, was not expected nor was it observed for these tests. The materials would, however, pyrolyze due to in-depth thermal decomposition. In addition, the LHMEL facility had a high throughput capability allowing for many tests at a low cost, which was essential for quick screening testing of many materials.

Testing consisted of irradiating a model with a CO_2 laser whose output irradiance was ~ 450 W/cm^2 for 60 seconds (Round 1) or 50 seconds (Round 2) to simulate an aerocapture maneuver. The laser spot size was 17 cm^2. Thermocouple, radiometer (for surface temperature measurements), and video data were recorded during each run. After a minimum of one day's cool down, one sample of each material was exposed to a second heat pulse of ~115 W/cm^2 for 100 seconds to simulate an entry environment. The models were photographed and weighed immediately before and after each test. Pre- and post-test height measurement, char depth, and photographs of sectioned models were also taken.

C. Description of Materials

A list of materials, vendors, test dates, and a brief description of the material are given in Table 1. For these Phase 1 (screening) tests, Round 1 testing occurred in May of 2010, and Round 2 of testing occurred in May of 2011.

Table 1. List of materials tested, vendors, dates, and a brief description of each material.

Name	Vendor	Dual Layer	Test Round	Description
PICA	Fiber Materials Inc (FMI)	no	1 & 2	Phenolic Impregnated Carbon Ablator
3DQP	Textron	yes	1	Three-layer system consisting of a dense outer mold, integration layer, and an insulation layer. Materials were quartz (silica) and phenolic.
BPA-FG	Boeing	yes	1	PICA-like material in honeycomb. "Functionally graded" transition layer between high- and low-density layers.
CC/Calcarb	Lockheed Martin Space Systems (LMSS)	yes	1	Top layer was a composite of carbon (graphite) in a carbon fiber reinforcement matrix. Bottom layer was a carbon foam made by Calcarb, Ltd.
MonA	LMSS	yes	1	A PICA-like "slurry" that has two different densities being stacked in the honeycomb and co-cured
PIRAS-22	Applied Research Associates (ARA)	no	1	A polyimide-based ablator system containing refractory reinforcing fibers and refractory fillers. The ablator has several reinforcement components one of which is refractory silicon-carbide fibers. To reduce final density it uses microballoon (MB) fillers including developmental silicon-carbide MBs and more common phenolic MBs
Phencarb P28/P15	ARA Laboratory	yes	1	PICA-like materials of different density packed atop one another
Graded PICA	FMI	yes	2	Variable density (graded) PICA
CBCF	FMI	no	2	Carbon bonded carbon fiber
Hoplon-22	ARA	no	2	A high-phenyl silicone ablator system containing refractory reinforcing fibers and a refractory filler. Both the fibers and microballoon fillers include silicon carbide materials of the same type as described for PIRAS-22
Resolite-18	ARA	no	2	Lightweight phenolic ablator

D. Model and Model Holder Design

All models were 5.08 cm (2.00 inch) in diameter and cylindrical in shape. The thickness of the PICA model was determined based on FIAT[12] predictions using the laser's irridation as an input energy term (multiplied by the surface absorptivity) in the surface energy balance. Ablation

(B') tables were not used, and ablation was not expected because the models were immersed in a nitrogen environment. The maximum allowable bond line temperature (see Fig.1) was 250°C and the back face of the FRCI-12 was adiabatic. The model surface was allowed to re-radiate to a surrounding environment at 21°C.

To have comparable test results between all materials tested, the mass of each model was requested to match the mass of the PICA model. Model length, therefore, varied for each material. Because each model had the same diameter, areal densities were therefore desired to be equivalent. It will be shown in the data analysis section, however, that areal density did, vary between materials depending on how well the vendors were able to match specifications.

1. Model Sizing for Round 1 Testing

FIAT predictions were made for a laser irradiation of 500 W/cm^2 for 180 seconds with a maximum bond line temperature of 250°C. The predicted PICA thickness was 6.35 cm (2.5 inch), with a resulting model mass of about 36 grams.

It was only much later after the test models had been made that it became known that the maximum laser duration for an intensity of 500 W/cm^2 was about 60 seconds. The decision was then made to drop the laser intensity to 450 W/cm^2 and irradiate the model for 60 seconds, which would be the new test conditions. However, with only about 1/3 of the desired (modeled) laser duration coupled with a loss in intensity of 50 W/cm^2, the test models would be severely oversized and bond line temperatures would not reach 250°C.

2. Model Sizing for Round 2 Testing

FIAT predicted a required PICA thickness of 3.40 cm (1.34 inch) when the sample was exposed to a laser intensity of 440 W/cm^2 for a duration of 50 seconds with a maximum bond line temperature of 250°C. The resulting model mass was about 18.93 grams. Round 2 testing of consisted of new PICA models and an entirely new set of vendor supplied materials.

3. Model Fabrication and Design

To help make the back wall of the material model adiabatic and to allow for thermocouple (TC) measurement, 1.27 cm (0.5 inch) of low density fibrous refractory composite insulation (FRCI-12)[13] was attached to the back of each model using RTV-560 adhesive.

Dual-layer models had two different materials stacked together, or two different densities of the same material stacked together. Typically, these models had a co-cured transition layer between them as shown in Figure 1. A listing of the models tested is given in Table 2.

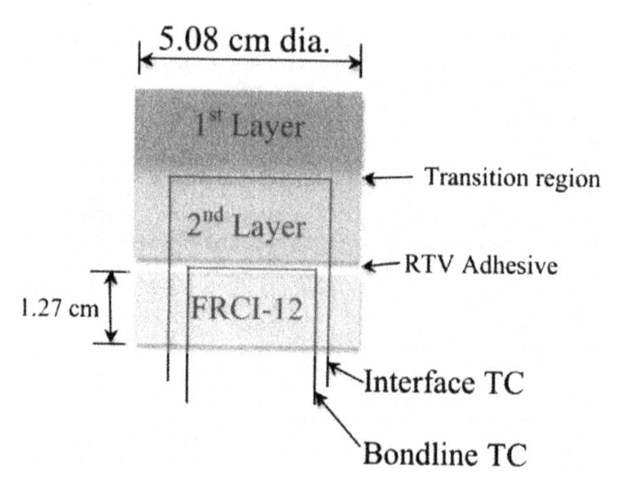

Figure 1. Assembly diagram of a typical dual-layer test model. Models with a single material of constant density did not have an interface TC.

4. Model Holder Design

The model holder is shown in Figure 2. The model (part #1) sat on three graphite pins (part #2) which were placed in slots on a graphite insert/sleeve (part #3). Pins of various diameter could be used to offset variations in model diameter. However, all models for these tests were within specifications of 5.08 cm (2.00 inch) diameter and were able to use the same size diameter pins, 0.635 cm (0.25 inch). The pins were incorporated into the holder design to minimize lateral heat conduction from the model to the graphite insert/sleeve. A Mullite block (part #4) housed the assembly and was placed on a moveable stage for laser beam alignment. The Mullite block was manufactured by Astro Met, Inc. of Cincinnati, Ohio. Finally, a back plate (part #5) was used as a stop for the graphite insert/sleeve.

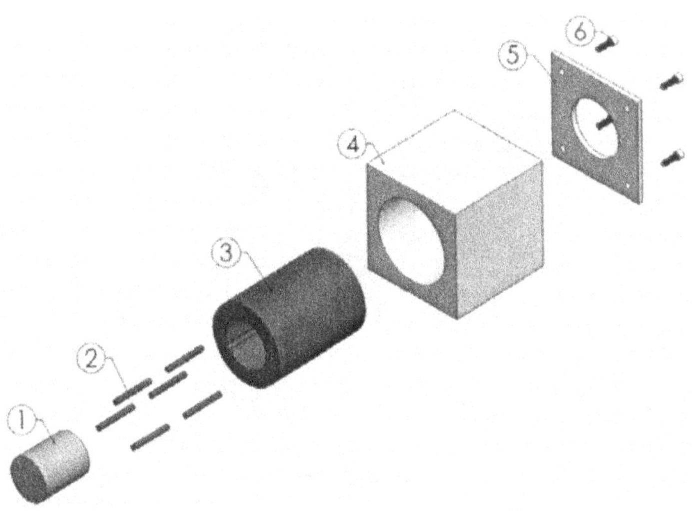

Figure 2. Assembly drawing of model holder.

E. Test Procedure

Test specimens were placed in the test holder, which was located inside the test box. To prevent oxidation and combustion of the test articles, a nitrogen feed line into the test box (see Figure 3) was added to ensure purging of oxygen from the system. In addition, the exhaust port on the test box was closed during, and for a few minutes after, each test run. It should be noted that for carbonaceous materials, these tests in nitrogen do not simulate ablative performance. The material response is only through conduction and pyrolysis.

Testing consisted of irradiating a model with the CO_2 laser at an intensity of approximately 450 W/cm^2 to simulate the heat flux in an aerocapture environment, with a spot size of 17 cm^2. The longest possible time that the laser could stay in continuous operation at this power was about 60 seconds, which was the test duration for Round 1 testing. To ensure a more consistent test duration, the test laser exposure time was reduced from 60 to 50 seconds for Round 2 testing. One sample of each material was exposed to a second heat pulse of ~115 W/cm^2 for 100 seconds to simulate the entry environment. There was at least one day between tests of the first and second heat pulse. Thermocouple, radiometer, and video data were recorded during each run. All models were photographed and weighed immediately before and after each test. Pre- and post-test height measurement, char depth, and photographs of sectioned models were also taken after the conclusion of all testing.

Figure 3. Plexiglass test box used to keep the test articles in a nitrogen-only environment

II. TEST RESULTS FROM ROUND 1

A. Pre- and Post-Test Photographs

1. PICA

Pre- and post-test photographs of the test specimens are shown in the following figures. All data are presented in the following section in Table 2. PICA test results were the benchmark to which test results from other models were compared. From Figure 4, it is apparent that PICA had a well-behaved pyrolysis and char response to the laser irradiation because the char layer exhibited no cracking, spallation, or pitting. Further, there was no evidence of swelling and the char layer was well defined.

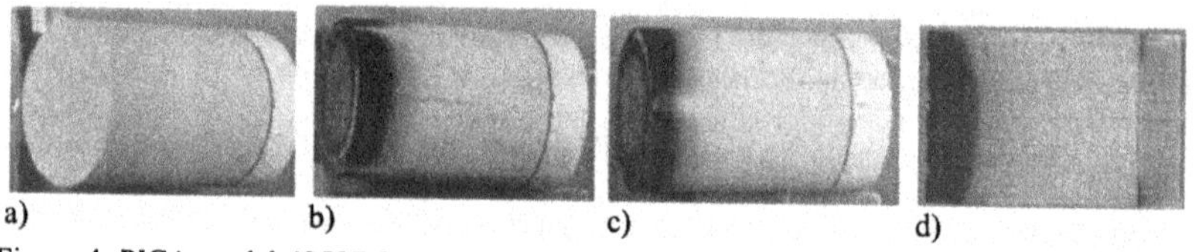

a) b) c) d)

Figure 4. PICA model 42527-3 a) pre-test b) post-test after 1st pulse c) post-test after 2nd pulse d) sectioned

2. 3DQP

Test results from Textron's dual layer material, 3DQP, are presented in Figure 5. For this test series, 3DQP had a severe adverse reaction beginning at the very instant of laser beam irradiation. Material continually came off of the model and filled the test chamber with debris. It was unknown why this material would behave in such an unexpected manner given its history of ablative performance. Two models were tested, each with unfavorable results. These samples were not exposed to the second, simulated entry heat pulse.

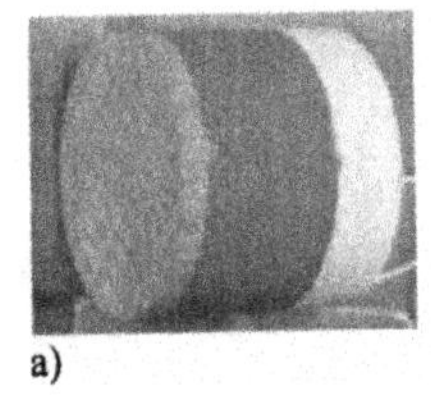
a)

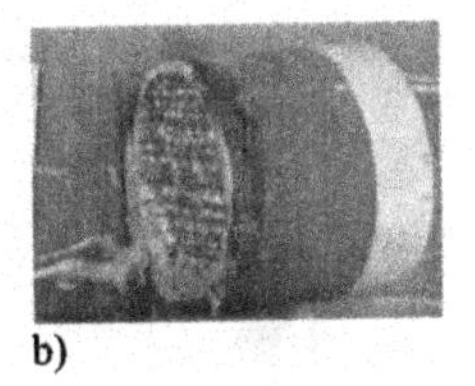
b)

Figure 5. 3DQP model A-02 a) pre-test and b) post-test photographs.

3. BPAFG

Figure 6 shows photographs from testing Boeing's functionally-graded, dual-layer, phenolic ablator, BPAFG. Overall, the material performed well, but it is apparent from these images (see circled region in Fig. 6a) that some of the low-density material adjacent to the honeycomb fell off during specimen manufacturing, possibly due to poor bonding to the honeycomb structure. The char layer was well-defined and the interface between layers has noticeable depth variations.

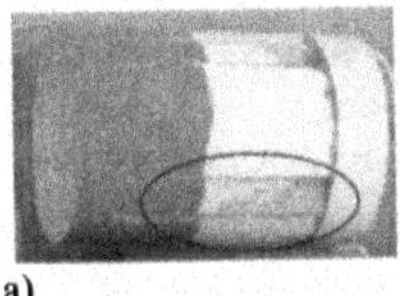
a)

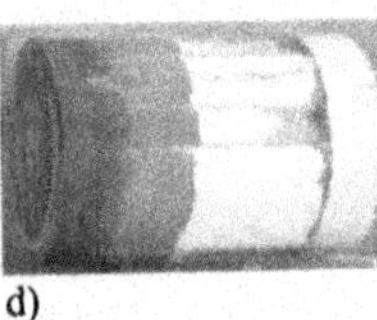
d)

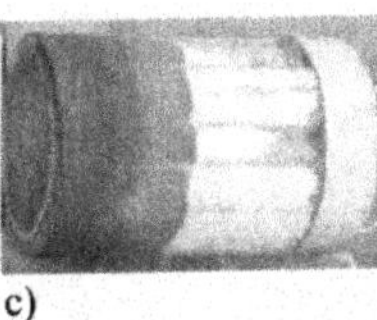
c)

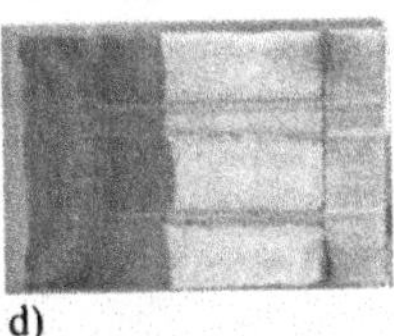
d)

Figure 6. BPAFG model A-03 a) pre-test b) post-test after 1st pulse c) post-test after 2nd pulse d) sectioned. The circled region of 6a) illustrates a region on the test sample where material had fallen off prior to testing.

4. CC/Calcarb

Figure 7 shows photographs of the test results from Lockheed Martin's Carbon-Carbon atop Calcarb dual layer model VF5000-01. Being a fully carbon material, no char layer is apparent in the figure. As expected, the carbon-carbon top layer shows very little effect from the laser irradiation.

a)

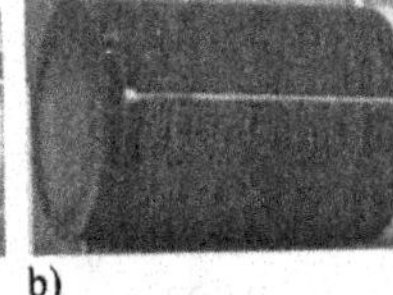
b)

c)

d)

Figure 7. CC/Calcarb model VF5000-01 a) pre-test b) post-test after 1st pulse c) post-test after 2nd pulse d) sectioned

5. MonA

Figure 8 shows photographs of the test results from Lockheed Martin's material MonA, model #A1. The images show MonA had a well-behaved pyrolysis and char response to the laser irradiation, with no evidence of swelling. Interestingly, as shown in sectioned model photograph (Fig. 8d), the char layer appears to have multiple layers based on distinct color bands. These bands were probably due to the differing levels of phenolic decomposition, and they appeared for all (single and dual laser pulse exposure) of the MonA tested models. The dual-layer interface also shows some depth variation, but this was not expected to affect the test results.

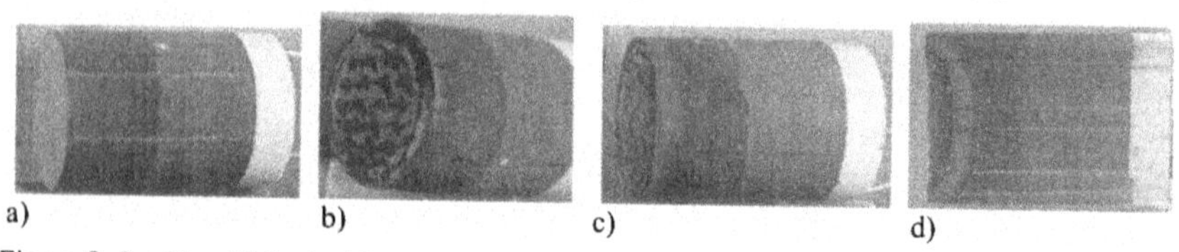

a) b) c) d)

Figure 8. Lockheed Martin MonA model A1 a) pre-test b) post-test after 1st pulse c) post-test after 2nd pulse d) sectioned

6. Phencarb P28/P15

Figure 9 shows photographs of the test results from ARA's Phencarb dual layer stackup material consisting of P28 over P15, model #9011. It was observed that along its outer surface, the dual layer interface was visually apparent (Fig 9a – 9d). The photographs show P28/P15 had a well-behaved pyrolysis and char response to the laser irradiation, with no evidence of swelling. However, the char layer appears to have some cracking as shown in sectioned model photograph (see circled region in Fig. 9d).

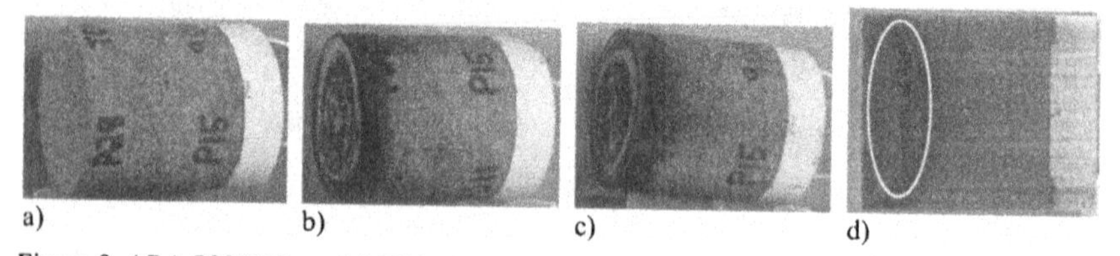

a) b) c) d)

Figure 9. ARA P28/P15 model 9011 a) pre-test b) post-test after 1st pulse c) post-test after 2nd pulse d) sectioned. The circled region highlights some cracking found in the char layer.

7. PIRAS-22

Figure 10 shows photographs of the test results from ARA's PIRAS-22, model #9011. The images show PIRAS-22 had a well-behaved pyrolysis and char response to the laser irradiation. However, the char layer did expand (swell) appreciably during the testing. The sectioned model photograph (Fig. 10d) shows the char layer to be well defined.

Swell was not considered to be a critical issue for material down-selection because to some extent it could help offset recession in an ablative environment. However, volume change (shrinkage or swelling) can cause issues such as cracking or shape irregularity as the piece size is increased.

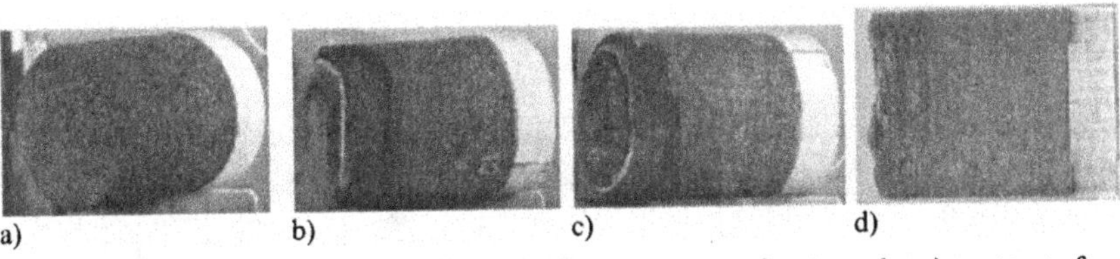

Figure 10. ARA PIRAS-22 model 9001 a) pre-test b) post-test after 1st pulse c) post-test after 2nd pulse d) sectioned

B. Summary of Round 1 Data

A summary of the data from Round 1 testing is given in Table 2. Results after a second laser exposure test are bolded and the model names are given the suffix "Char" to indicate that the model had previously been charred by exposure to the first laser pulse.

1. Surface recession

The models were tested in a nitrogen-rich environment to avoid the effects of ablation. However, surface recession (shrinkage) and char swelling occurred for some models. Using a depth gauge, post-test height measurements were averaged from four radial, evenly spaced measurement locations on the top of the model. These measurements were taken at NASA Ames after the completion of testing. The results are given in Table 2. Figures 11 and 12 show the averaged results from the first laser pulse (simulating aerocapture) and the averaged results for the second laser pulse (simulating entry). The amount of recession was determined by the difference in pre- and post-test height measurements. The accuracy of height and recession measurements was ± 0.025 cm (0.010 inch) based on repeatability of the measurements.

Height measurements were taken before testing and after a model's final laser exposure (first or second pulse). Consequently, initial height and recession after the first laser exposure data are not reported for the dual-pulse models (see Table 2).

PICA, Phencarb P28/P15, and CC/Calcarb did not experience any surface recession. 3DQP experienced catastrophic behavior with what appeared to be virgin material coming off the surface. MonA and BPAFG did experience a small amount of shrinkage ~ 0.1 cm (0.04 in). The only material to show significant swell was PIRAS-22 at approximately 0.7 cm (0.276 in.), which was easily noticeable when the tested models were observed.

Table 2. Summary of Round 1 test data. Data in bold and marked with the suffix "Char" are the test results for models that had previously been exposed to the first laser conditions (to simulate an aerocapture maneuver) and undergone a second laser exposure (to simulate planetary entry).

Name	Laser Duration, s	Laser Intensity, W/cm²	Heat Load, kJ/cm³	Max. Interface Temp, °C	Max. Bondline Temp, °C	%Virgin Remaining	Recession, cm	Initial Model Mass, g	Mass Loss, g
PICA 42527-3	60.1	447	26.8	n/a	41	-	-	51.197	1.51
PICA 42527-3 Char	**100.0**	**117**	**11.7**	**n/a**	**45**	**72.9**	**0.011**	**49.928**	**0.15**
PICA 42527-4	57.6	453	26.1	n/a	46	73.1	0.007	51.567	1.28
3DQP-A-01	46.5	448	20.8	2845	71	59.8	0.350	62.129	12.43
3DQP-A-02	28.8	459	13.2	n/a	56	75.1	0.350	63.362	9.90
BPAFG-A-01	52.8	437	23.1	75	49	67.9	0.182	65.124	4.61
BPAFG-A-02	56.9	444	25.3	76	46	67.1	0.122	66.304	3.62
BPAFG-A-03	57.5	453	26.1	74	49	-	-	65.318	5.37
BPAFG-A-03 Char	**100.1**	**114**	**11.4**	**83**	**46**	**58.5**	**0.158**	**60.190**	**1.31**
CC/CALCARB VF5000-01	57.0	448	25.5	1991	75	-		64.352	0.28
CC/CALCARB VF-5000-01 Char	**100.1**	**119**	**11.9**	**1322**	**55**	**100.0**	**-0.016**	**64.155**	**0.09**
CC/CALCARB VF-5000-02	57.6	442	25.4	2069	60	100.0	0.008	62.749	0.35
CC/CALCARB VF-5000-03	57.6	454	26.1	2004	60	100.0	-0.010	63.048	0.26
LM-MONA-A1	57.6	445	25.6	77	51	-	-	62.782	3.19
LM-MONA-A1 Char	**100.1**	**114**	**11.4**	**87**	**43**	**69.9**	**0.129**	**59.829**	**0.65**
LM-MONA-A2	57.5	453	26.0	77	53	73.0	0.116	62.575	2.98
LM-MONA-A4	57.5	451	25.9	81	48	72.1	0.125	62.707	3.27
P28/P15-9011	57.6	444	25.6	67	39	-	-	61.106	4.13
P28/P15-9011 Char	**100.0**	**116**	**11.6**	**128**	**47**	**67.7**	**-0.009**	**57.224**	**1.38**
P28/P15-9012	53.6	433	23.2	56	36	74.4	-0.030	62.757	3.77
P28/P15-9013	57.6	448	25.8	63	39	76.6	-0.003	61.280	4.14
PIRAS-22 9001	60.1	431	25.9	n/a	40	-	-	51.831	3.62
PIRAS-22-9001 Char	**100.1**	**116**	**11.6**	**n/a**	**41**	**77.9**	**-0.683**	**48.448**	**0.36**
PIRAS-22-9002	50.4	435	21.9	n/a	40	80.2	-0.706	52.631	2.93
PIRAS-22-9003	57.6	446	25.7	n/a	36	80.0	-0.721	51.785	3.24

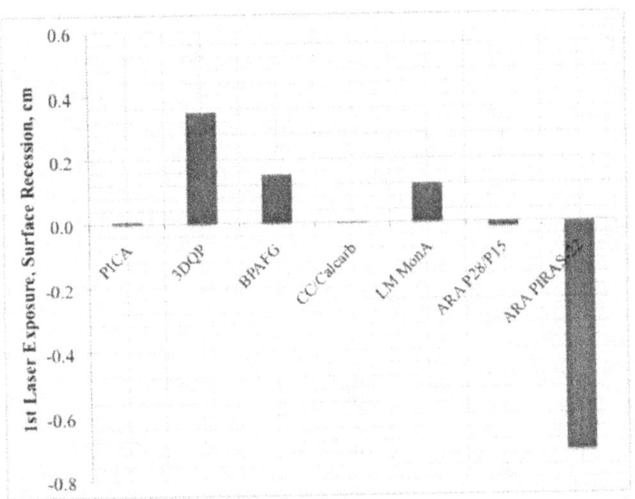

Figure 11. Surface recession of single laser pulse tested materials. Negative values indicate surface expansion (swell).

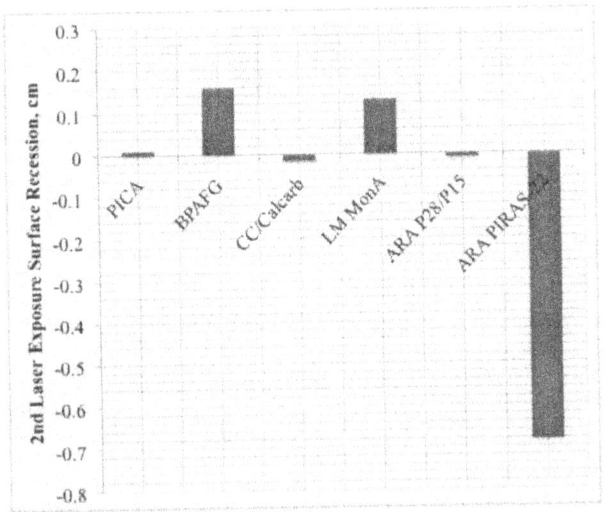

Figure 12. Surface recession for Round 1 tested materials after the second laser exposure. Negative values indicate surface expansion (swell).

2. Areal Density

For comparable test results, the mass of each model was requested to match the PICA model mass of approximately 36 grams. Because each model had the same diameter, 5.08 cm, model thickness varied to keep the areal density (and mass) constant. Areal densities were calculated by dividing initial model mass (not including FRCI-12 back plate or TC) by surface area (20.268 cm^2). A comparison of the averaged areal density for each virgin material is given in Figure 13. Of the materials tested, BPAFG had the highest areal density, 1.94 g/cm^2, which was about 11% greater than PICA. The dispersion in areal density between materials exists due to how well each vendor was able to meet the design specification for model mass to be 36 grams.

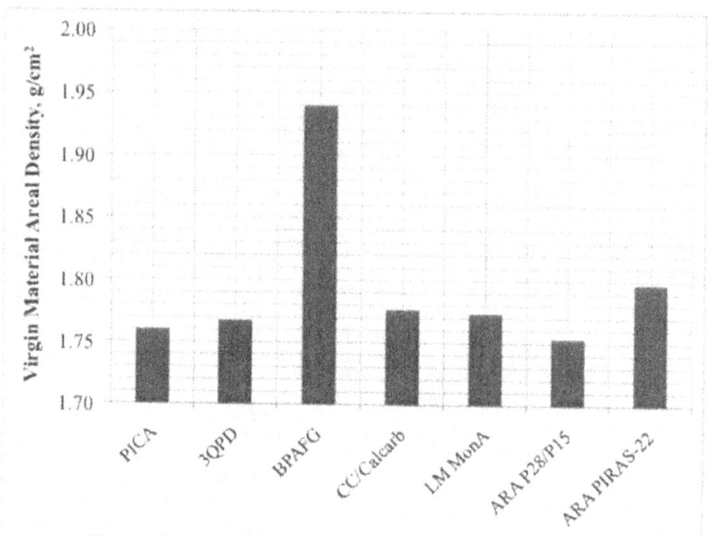

Figure 13. Averaged areal density of Round 1 tested materials

3. Mass loss

The total initial model mass (including FRCI-12 back plate TC wires) and mass loss data are given in Table 2. Mass measurements were taken immediately before and after each test because atmospheric water absorption would allow model mass to vary. The results are shown in Figure 14 for the first laser exposure and Figure 15 for the second laser exposure. Atmospheric water absorption was especially evident with post-tested models, whose mass was observed to slowly drift upwards for an hour or more after being tested. It was for this reason that mass loss data were reported with less accuracy (significant figures).

As the testing baseline, PICA, when tested from its virgin state, had an average mass loss of 1.395 grams, and the charred PICA sample (second laser pulse) lost 0.15 grams. All other materials tested, with the exception of Carbon-carbon/Calcarb, had greater mass loss for both virgin and charred (second-pulse) samples. The data are given in Table 2. CC/Calcarb had an average mass loss of 0.30 grams for the virgin material and 0.09 grams for the charred sample, second-pulse test.

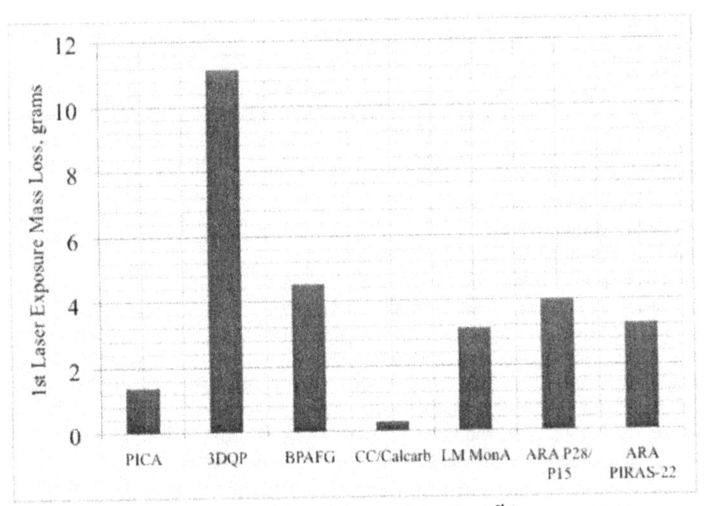

Figure 14. Mass loss of Round 1 materials after 1st laser exposure

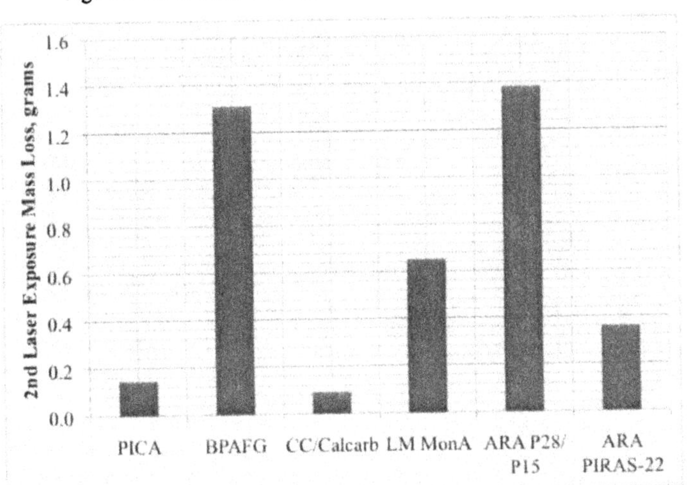

Figure 15. Mass loss of Round 1 materials after 2nd laser exposure (note: 3DQP was not tested for a second laser exposure)

4. Percentage virgin material remaining

One way of looking at char depth was by percentage of virgin material remaining. Using a sample sectioned along the cylinder axis, percentage virgin material remaining was determined by measuring, based on visual inspection, the distance from the bond line to the beginning of pyrolysis edge and then dividing this length by the original model thickness. This was a subjective test because the exact location of the beginning of pyrolysis edge can only be determined by knowing the local density. The pyrolysis edge is usually defined as the location where the local density is equivalent to 98% of the density of virgin material. Nevertheless, this test may give a qualitative indication of the depth of thermal decomposition for a material. The results are given in Table 2.

Figure 16 shows the results for all of the models. Values are given as a percentage because model thickness varied between materials and model thickness was determined by the requirement that each material had an equivalent areal density.

As the benchmark material, PICA had values of 73.1% for the virgin material tested and 72.9% for the charred. Being a carbon-only material, CC/Calcarb did not exhibit any pyrolysis zone and thus had 100% of virgin material remaining for all models tested. PIRAS-22 had the next highest values at 80.1% for virgin material tested, and 77.9% for charred. Phencarb P28/P15 had values slightly better than PICA for the virgin material tested at 75.5% and slightly worse for the charred material at 67.7%. MonA had values similar to PICA for virgin material tested and slightly worse at 69.9% for charred material. BPAFG had worse results than PICA, with values of 67.5% for virgin material tested and 58.5% for charred material tested.

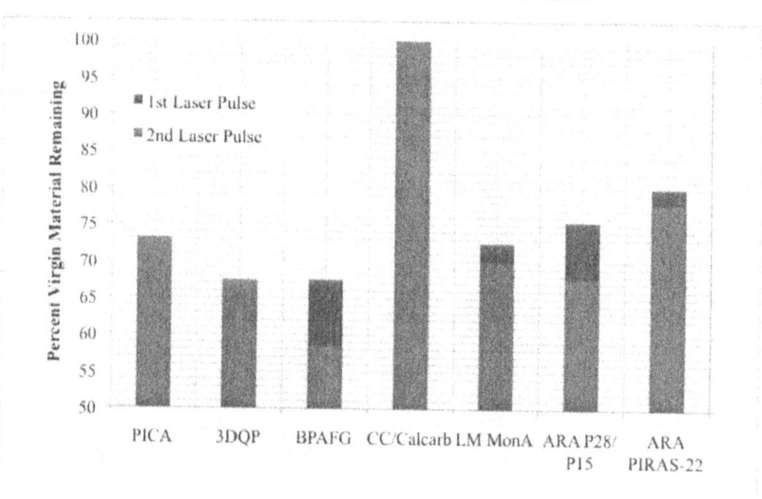

Figure 16. Estimated percentage virgin material remaining for Round 1 tested materials

5. Maximum bond line temperature

Maximum bond line temperatures for all models tested are given in Table 2. Figure 17 shows a comparison of the averaged maximum bond line TC temperature recorded for the models after the first laser pulse exposure. Although the laser duration was for the first exposure was set to 60 seconds, the laser would often shut down a few seconds prior to this (see Table 2). Because heat load had such a strong influence on bond line temperature, the material's averaged heat load [kJ/cm^2] for the tests is given above each bar. With the exception of 3DQP, for which the laser had to be manually shut off early, the average heat load varied by no more than 6.3% for each material, and average laser intensity varied by no more than 6.4%. As shown in Fig. 16, only ARA's P28/P15 and ARA's PIRAS-22 were able to achieve a lower average bond line temperature. Note that the maximum bond line temperature from the second laser pulse had a span of only 16°C, with most of the materials within 6°C. Given this narrow range, no further data analysis was performed.

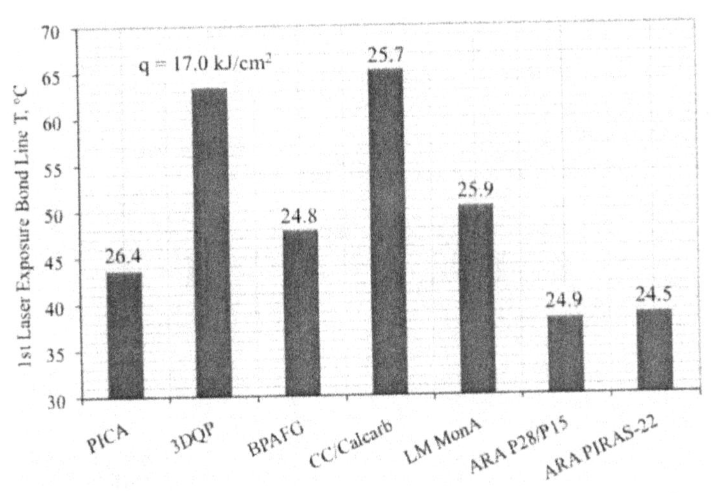

Figure 17. Averaged maximum bond line temperature and averaged heat load [kJ/cm^2] for models after being exposed to the first laser pulse.

III. DISCUSSION OF ROUND 1 TEST RESULTS

Of all the materials tested, ARA PIRAS-22 had the best performance when compared to PICA. It had similar values for maximum bond line temperature and slightly higher values for mass loss and areal density. PIRAS-22 outperformed PICA for percentage virgin material remaining. Interestingly, the material swelled when exposed to the laser pulse, resulting in an increase in surface height above the untested model. While PICA recessed, PIRAS-22 did not.

BPAFG had greater surface recession, areal density, and mass loss than PICA, but had similar values of maximum bond line temperature. BPAFG also had a worse percentage virgin material remaining especially after the second laser pulse.

LM MonA had higher surface recession and greater mass loss than PICA, but had similar areal density, bond line temperature, and percentage virgin material remaining.

ARA P28/P15 had a slight swell to it (no recession) after laser exposure. The material had better maximum bond line temperatures after the first laser pulse exposure, and similar values after the second laser pulse exposure. ARA P28/P15 had similar values of areal density and percentage virgin material remaining, but had a greater mass loss when compared to PICA.

As expected, CC/Calcarb remained in its "virgin" state throughout the testing, resulting in a value of 100% for percentage virgin remaining. The material had very little mass loss and no recession, but also had the greatest values of maximum bond line temperature. Its areal density was similar to PICA's.

3DQP had poor results.

Results from these tests were used as part of the down-selection process for determining which materials would move into the EDL TDP development phase and which materials would remain in the screening process.

IV. TEST RESULTS FROM ROUND 2

A. Pre- and Post-Test Photographs

1. PICA

The results for PICA are identical in appearance to those found in Round 1 testing (see Fig. 4). The results from Round 2 testing are shown in Figure 18.

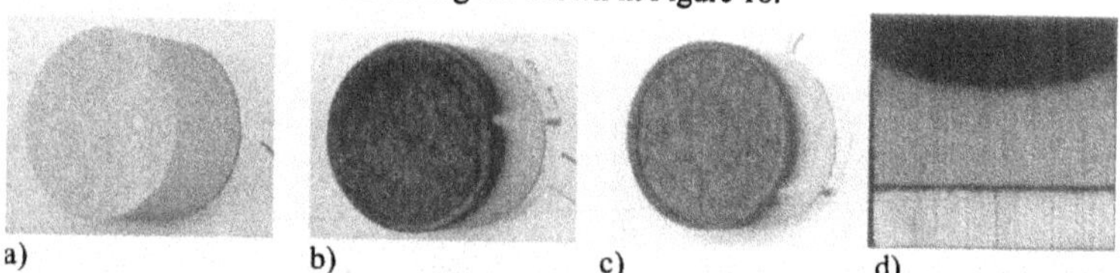

a) b) c) d)

Figure 18. PICA model L-RR a) pre-test b) post-test after 1st pulse c) post-test after 2nd pulse d) sectioned

2. Round 2 vendor materials

The results for Graded PICA, Hoplon-22, Resolite-18, and CBCF are shown in Figures 19-22.

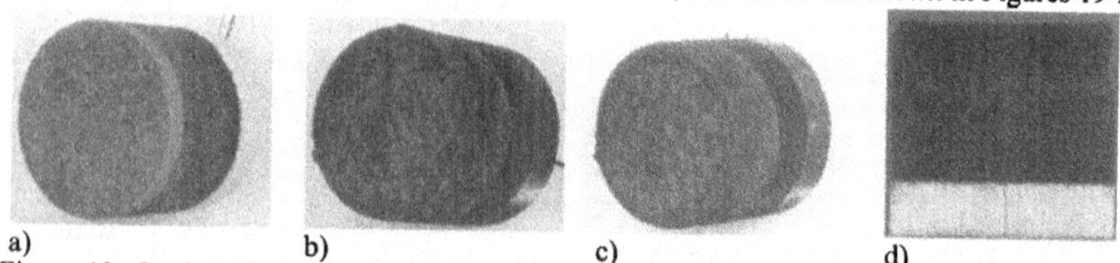

a) b) c) d)

Figure 19. Graded PICA model 01 a) pre-test b) post-test after 1st pulse c) post-test after 2nd pulse d) sectioned

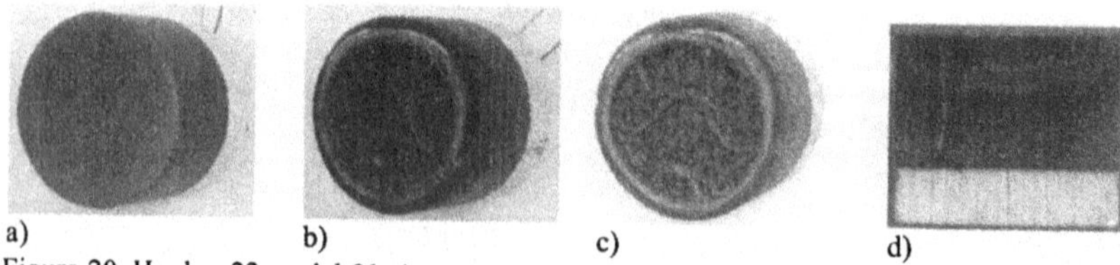

a) b) c) d)

Figure 20. Hoplon-22 model 01 a) pre-test b) post-test after 1st pulse c) post-test after 2nd pulse d) sectioned

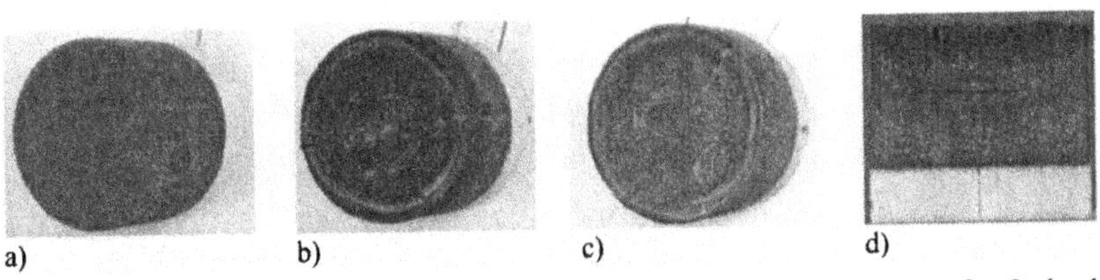

Figure 21. Resolite-18 model 02 a) pre-test b) post-test after 1st pulse c) post-test after 2nd pulse d) sectioned

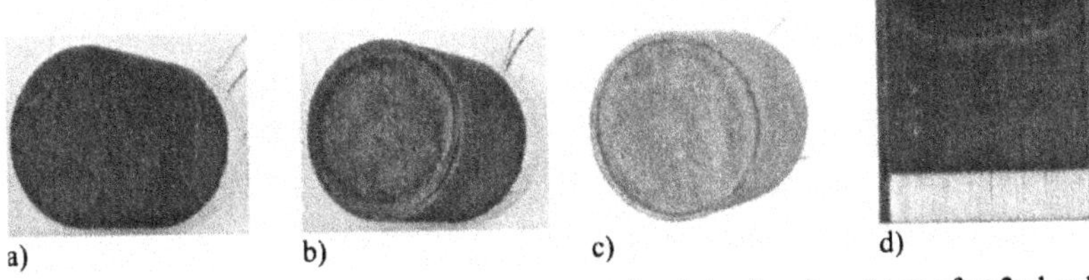

Figure 22. CBCF model 02 a) pre-test b) post-test after 1st pulse c) post-test after 2nd pulse d) sectioned

B. Summary of Round 2 Data

A summary of the data from Round 2 testing is given Table 3. Results after a second laser exposure test are bolded and the model names are given the suffix "Char" to indicate that the model had previously been charred by exposure to the first laser pulse.

Table 3. Summary of Round 2 test data. Data marked with the suffix "Char" are test results for models that had previously been exposed for 50 seconds to the first laser conditions (to simulate an aerocapture maneuver) and undergone a second laser exposure for 100 seconds (to simulate planetary entry).

Name	Laser Intensity, W/cm²	Max Bondline Temp, °C	Initial Mass, g	Areal Density g/cm²	Mass Loss, g	Heatload, kJ/cm²	%Virgin Remaining
PICA_1	449.4	95.3	32.249	0.955	1.279	22.5	
PICA_1 Char	116.3	107.4	30.970		0.102	11.6	60.6
PICA_2	453.8	91.2	32.674	0.958	1.244	22.7	60.7
Graded PICA_01	444.0	384.0	27.615	0.730	2.625	22.2	
Graded PICA_01 Char	116.9	151.5	24.990		0.056	11.7	42.7
Graded PICA_02	454.9	150.1	27.533	0.734	1.087	22.8	53.1
Hoplon-22_01	449.4	88.7	33.784	1.030	4.531	22.5	
Hoplon-22_01 Char	116.0	190.8	29.253		1.046	11.6	27.8
Hoplon-22_02	450.2	91.5	34.139	1.042	4.438	22.6	35.9
Resolite-18_02	455.4	85.2	30.467	0.886	3.699	22.8	
Resolite-18_02 Char	114.9	115.3	26.768		0.557	11.5	29.2
Resolite-18_03	451.2	85.1	31.673	0.926	3.661	22.6	44.7
CBCF_01	449.0	65.9	31.944	0.920	0.849	22.4	
CBCF_01 Char	115.7	58.8	31.095		0.174	11.6	60.7
CBCF_02	450.5	63.5	31.479	0.925	0.939	22.6	61.6

1. Surface recession

The models were tested in a nitrogen-rich environment to avoid the effects of ablation. Using vernier calipers, pre- and post-test thickness measurements were made, and recession was defined as the difference between initial and final thickness values. The accuracy of these measurements was ± 1.0 mm (0.039 inch) based on repeatability and softness of the char to the caliper. Relative to this degree of accuracy, none of the models showed any significant shrinkage (recession) or swell.

2. Areal Density

The areal density for each model prior to testing is given in Table 3 and shown in Figure 23. Graded PICA, with a value of ~0.73 g/cm^2, was substantially lower than PICA (0.96 g/cm^2). Resolite-18 and CBCF have values similar to PICA, while Hoplon-22's areal density was greater (1.04 g/cm^2). Because the desired mass of PICA for each round of testing was different, the values of areal density for PICA are also different between the two rounds.

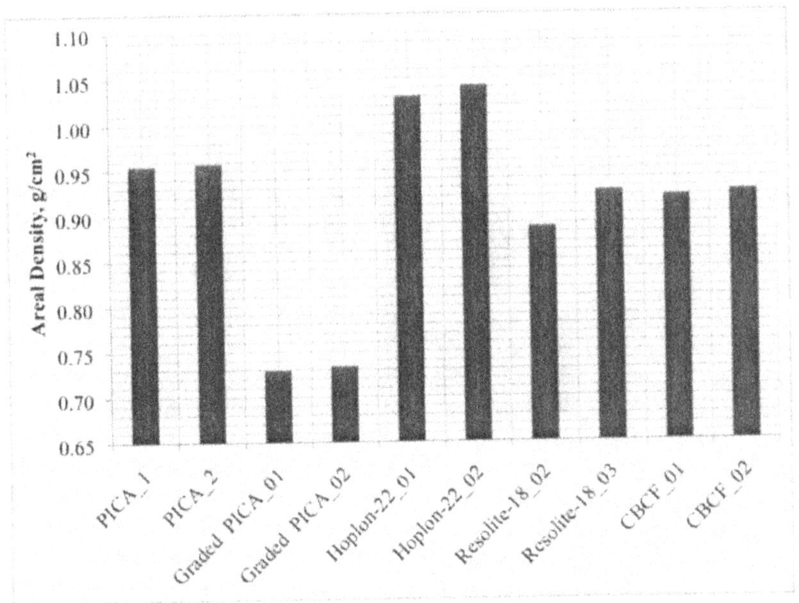

Figure 23. Areal Density (grams/cm^2) for Round 2 virgin materials

3. Mass loss

Mass loss data are in Table 3. All models had a TC located at the bond line between the model and FRCI-12 insulator. Measurements were taken immediately before and after each test because atmospheric water absorption could allow model mass to vary. The results after the first laser exposure are shown in Figure 24 and results after the second laser exposure are shown in Figure 25.

As the testing baseline, PICA, when tested from its virgin state, had an average mass loss of 1.261 grams, and the charred PICA sample, when exposed to the second laser pulse, lost 0.102 grams.

For the first laser pulse exposure, FMI's CBCF and Graded PICA (model# 02) had a smaller mass loss, and ARA's Hoplon-22 and Resolite-18 had greater (see Fig. 25). For the second laser exposure, charred CBCF had results similar to PICA, Graded PICA (model# 01) had lower values, while both Resolite-18 and Hoplon-22 continued to show greater mass loss than PICA.

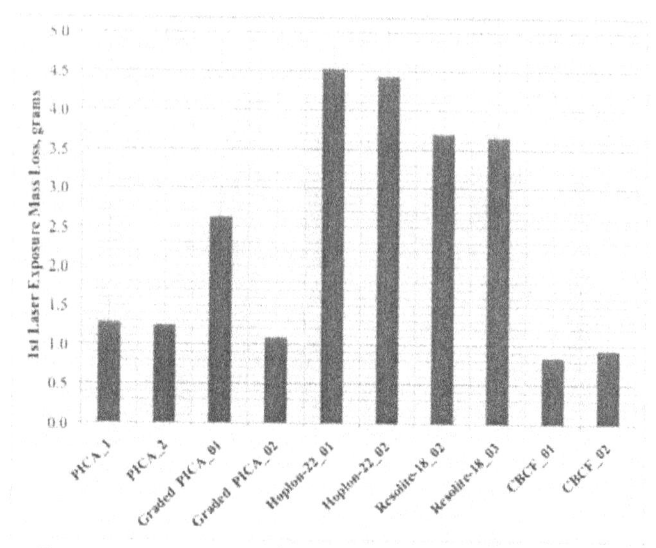

Figure 24. Mass loss from 1st laser exposure of Round 2 materials

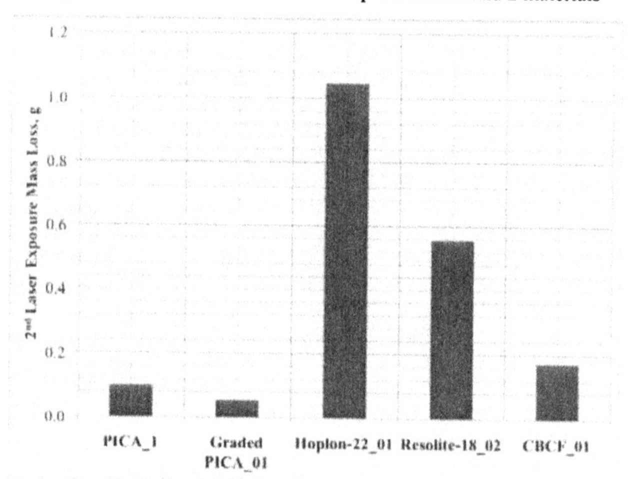

Figure 25. Mass loss after second laser exposure for Round 2 tested materials

4. Percentage virgin material remaining

Using a sectioned sample, the percentage of virgin material remaining was found by measuring the distance from the bond line to the beginning pyrolysis edge, and then dividing this length by the original model thickness. Model thickness varied between materials because of the requirement that each material had an equivalent areal density. The results are given in Table 3 and shown in Figure 26.

As the benchmark material, PICA had values of 60.71% virgin material remaining after the first laser pulse and 60.57% after the second laser pulse. For this metric, only CBCF out-performed PICA, with values of 60.61% and 60.67% respectively. Hoplon-22 had the lowest values of 35.39% and 27.78%, respectively.

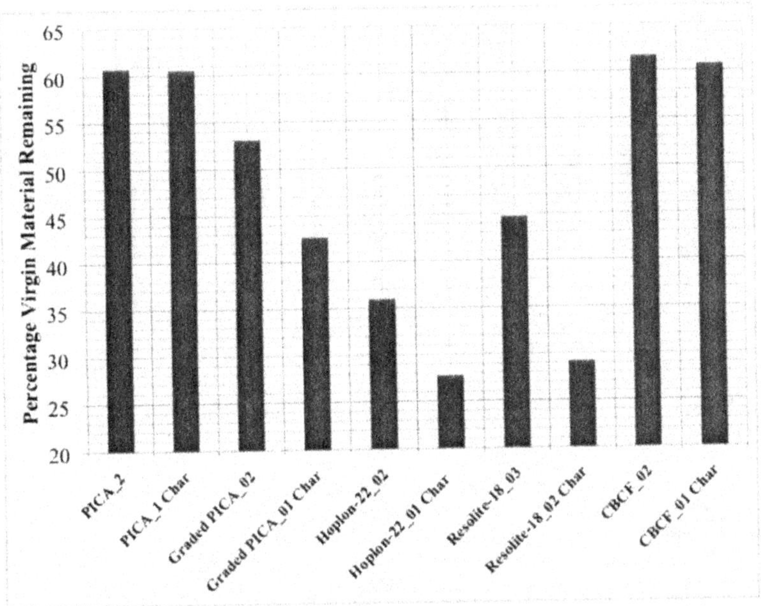

Figure 26. Percentage virgin material remaining for Round 2 tested materials

5. Maximum bond line temperature

Maximum bond line temperatures for all models tested are given in Table 3. Figure 27 shows a comparison of the maximum bond line TC temperatures recorded after the first laser pulse exposure, and Figure 28 shows a comparison of the maximum bond line TC temperatures recorded after the second laser exposure.

For the results from first laser exposure, the benchmark material, PICA, had a maximum bond line temperature of 93.2°C. The only material to substantially outperform PICA for this metric was CBCF, with a value of 64.7°C. The material with the highest values was Graded PICA (model# 01), with an exceptionally high value of 384°C. This value is probably due to having the fiberform act to channel the pyrolysis gases to the thermocouple, thereby measuring the pyrolysis gas temperature. For the second Graded PICA model tested, the maximum bond line temperature was a more reasonable 150.1°C. Based on the results shown in Fig. 27, Hoplon-22 and Resolite-18 had very similar responses compared to virgin PICA.

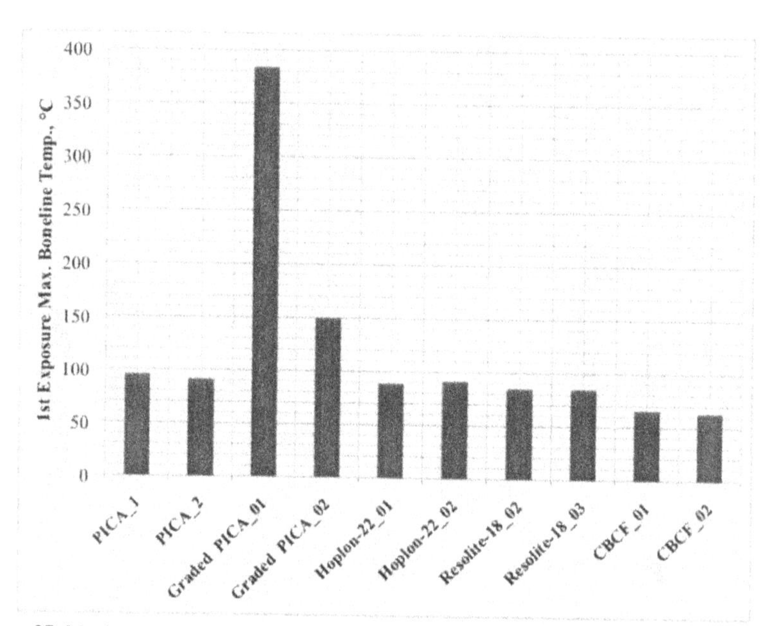

Figure 27. Maximum bond line temperature for Round 2 models after being exposed to the first laser pulse

The maximum bond line temperature results from the second laser pulse (representing vehicle entry) on the charred samples are shown in Figure 28. Here, Graded PICA (151°C) and Hoplon-22 (191°C) had values much higher than PICA (107°C), while Resolite-18 was similar (115°C) and CBCF was much lower (59°C). These results are possibly related to mass loss, and a correlation seems to exist between the amount of mass lost from the first laser pulse relative to PICA (Fig. 25) and maximum bond line temperature from the second pulse. Generally speaking, if a material lost more mass than PICA in the first laser exposure, its bond line temperature would increase over that of PICA during the second laser exposure.

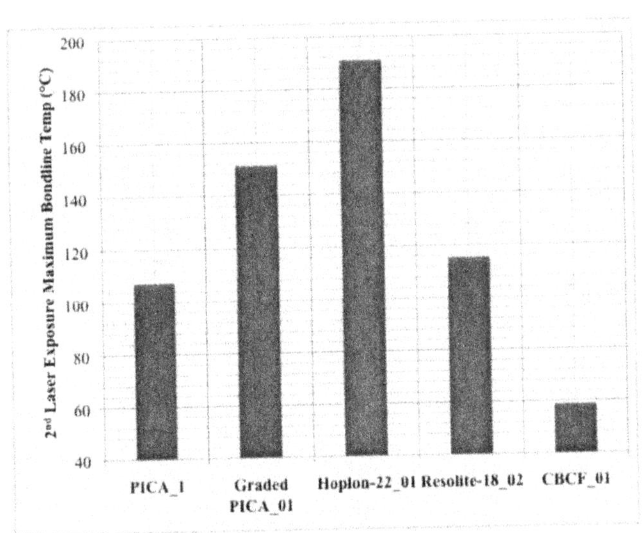

Figure 28. Maximum bond line temperature for Round 2 models after being exposed to the second laser pulse.

V. DISCUSSION OF ROUND 2 TEST RESULTS

Of all the materials tested, CBCF had the best performance when compared to PICA. It had a lower maximum bond line temperature for each laser exposure and similar values for percentage virgin material remaining, mass loss, and areal density.

Graded PICA had a much lower areal density than PICA. Of the two models tested, one (model# 01) had substantially higher bond line temperatures and mass loss compared to PICA or the other Graded PICA model (model# 02). It is unclear why this model had these results. Graded PICA model# 02 had higher bondline temperatures for both laser pulses, lower percentage virgin material remaining, and similar mass loss when compared to PICA

Hoplon-22 had similar maximum bondline temperatures compared with PICA for the first laser pulse, but for the second laser pulse, its value was much greater. Hoplon-22 also had a greater char depth (lower percentage of virgin material remaining) and greater mass loss and areal density when compared to PICA.

Resolite-18 had similar maximum bond line temperatures for the first laser exposure compared to PICA, but much higher temperatures for the second laser pulse. It had greater mass loss and a lower percentage of virgin material remaining than PICA, but its areal density was similar to PICA. Some cracks are evident in the sectioned sample (see Fig. 21d).

Results from these tests were used as part of the down-selection process for determining which materials would move into the EDL TDP development phase and which materials would remain in the screening process.

VI. CONCLUSIONS

Novel materials have been developed and supplied for testing at Wright-Patterson Air Force Base's LHMEL testing facility as part of the NASA Entry, Descent, and Landing Project. The test objectives of exposing the materials to a CO_2 laser pulse and recording bond line temperatures, mass loss, and char depth (as a percentage of virgin material remaining) for this work have been completed. Pre- and post-test photographs of the models have also been taken.

One of the goals of the program was to demonstrate the ability of new materials to be made of varying density through the material thickness. This capability may allow a heat shield designer to tailor the density of a material to take advantage of having a higher density for ablation and a lower density for an insulator. For Round 1 of testing, nearly all of the materials except PICA and PIRAS-22 had a "dual layer" design with different density ablators stacked atop one another. For Round 2 testing, Graded PICA was the only material of varying density.

Another goal of the project was to demonstrate the ability to place PICA-like materials and materials of varying density into honeycomb. Being able to pack a material in honeycomb has the advantage of being able to produce one type of monolithic heat shield. For the Round 1 tested materials, MonA and BPAFG were each packed in honeycomb.

For the test metric of char depth (given as a percentage material remaining), the Lockheed Martin CC/Calcarb tested in Round 1 showed no charring (because the virgin material is only carbon) while the FMI CBCF tested in Round 2 showed results comparable to PICA.

Another test metric was maximum bond line temperature. ARA's P28/P15 and PIRAS-22 tested in Round 1 were both able to outperform PICA. For materials tested in Round 2, FMI's CBCF consistently *outperformed* PICA, while ARA's Resolite-18 had results similar to PICA.

This work was part of the NASA EDL TDP TPS element for testing of candidate, vendor-supplied, rigid TPS materials. The results generated from this testing program will be used as part of the down-selection process to determine which materials will continue to the next stage of development and which materials will need to repeat the screening process. Material development activities in the next stage includes determining material properties, developing material response models, and performing arc-jet testing for material response.

VI. ACKNOWLEDGMENTS

The authors would like to acknowledge the support of the Exploration Technology Development and Demonstration (ETDD) Program, managed at NASA Glenn Research Center. The work documented in this paper was performed as part of ETDD Entry, Descent, and Landing (EDL) Technology Development Project, which is managed at NASA Langley Research Center and supported by NASA Ames Research Center, NASA Johnson Space Center, and the Jet Propulsion Laboratory.

The authors also gratefully acknowledge the support provided by the Thermal Protection Materials Branch and the Aerothermodynamics Branch of NASA Ames Research Center through NASA Contract No. NNA10DE12C with ERC Corporation.

VII. REFERENCES

[1] Edquist, K.T., Dyakonov, A., Wright, M.J., and Tang, C.Y. "Aerothermodynamic Design of the Mars Science Laboratory Heatshield" AIAA paper 2009-4075, 41st AIAA Thermophysics Conference, June 2009, San Antonio, Texas.
[2] Tran, H., Johnson, C., Rasky, D., Hui, F., Hsu, M., Chen, Y-K., "Phenolic Impregnated Carbon Ablators (PICA) For Discovery Class Mission", AIAA paper 1996-1911, presented at 31st AIAA Thermophysics Conference, New Orleans, LA, June, 1996.
[3] Henline, W.D., "Thermal Protection Analysis of Mars-Earth Return Vehicles," Journal of Spacecraft and Rockets, Vol. 29, No. 2, March-April 1992.
[4] Wright, M., Beck, R.., Slimko, E., Edquist, K., Driver, D., Sepka, S., Slimko, E., Willcockson, W., DeCaro, A., "Sizing and Margins Assessment of the Mars Science Laboratory Aeroshell Thermal Protection System", AIAA paper 2009-4231 41st AIAA Thermophysics Conference, June 2009, San Antonio, Texas.
[5] Kontinos, D., Stackpoole, M., "Post-Flight Analysis of the Stardust Sample Return Capsule Earth Entry,"AIAA 2008-1197, January 2008.
[6] Beck, R., Driver, D., Wright, M., Laub, B., Hwang, H., Sepka, S., Slimko, E., Edquist, K., Willcockson, W., Thames, T., "Development of the Mars Science Laboratory Heatshield Thermal Protection System", AIAA paper 2009-4229 41st AIAA Thermophysics Conference, June 2009, San Antonio, Texas.
[7] McGuire. M.K., "Dual Heat Pulse, Dual Layer Thermal Protection System Sizing Analysis and Trade Studies for Human Mars Entry Descent and Landing", AIAA paper 2011-343, presented at 49th AIAA Aerospace Sciences Meeting including the New Horizons Forum and Aerospace Exposition 4 - 7 January 2011, Orlando, Florida
[8] Arnold, J., Venkatapathy, E., Chen, Y-K.,Sepka, S., Argrwal, P., "Validation Testing of a New Dual Heat Pulse, Dual Layer Thermal Protection System Applicable to Human Mars Entry, Descent and Landing", AIAA paper 2010-5050, presented at 10th AIAA/ASME Joint Thermophysics and Heat Transfer Conference 28 June - 1 July 2010, Chicago, Illinois
[9] Laub, B., Venkatapathy, E., "Thermal Protection System Technology and Facility Needs for Demanding Future Planetary Missions", presented at the International Workshop on Planetary Probe Atmospheric Entry and Descent Trajectory Analysis and Science, Lisbon, Portugal, 6-9 October 2003
[10] Zell, P., Venkatapathy, E., Arnold, J., "The Block-Ablator-In-a-Honeycom Heat Shield Architecture Overview", presented at the International Planetary Probe Workshop 7, Barcelona, Spain 12-18, June 2010.
[11] Eric, J., "Laser Hardened Materials Evaluation Laboratory Performs Evaluation Of Advanced Materials For Aerospace Systems", ML-00-15, AFRL Technology Horizons, Volume 1, Number 4, page 10, Dec. 2000.
[12] Chen, YK., Milos, F.S., "Ablation and Thermal Response Program for Spacecraft Heatshield Analysis," Journal of Spacecraft and Rockets, Volume 36, Number 3, May-June 1999
[13] Leiser, D.B., Goldstein, H. E., and Smith, M., "Fibrous Refractory Composite Insulation," U.S. Patent 4, 148, 962, 1978.

FAILURE MECHANISM OF R-TYPE TEMPERATURE SENSOR IN EXTREME ENVIRONMENTS

Hongy Lin and Eric Allain
Watlow Electric Manufacturing Co.
St Louis, MO, USA

ABSTRACT

The R-type thermocouple has long been used in high-temperature processing environments for providing a feedback signal for temperature control. Current advancements in the semiconductor industry require a non-oxidizing manufacturing environment and process that generates elemental contaminants, severely impacting the life of the temperature sensor. The environment is characterized by significant Si vapor pressure, while a complex atmosphere containing Mg, Al, Si, etc., is present inside the Al_2O_3 protection tube where the thermocouple resides. The high temperature (> 1500°C) and corrosive environment often leads to premature thermocouple failure, which jeopardizes the quality of the products. The primary failure mode of an R-type thermocouple is Pt wire breakage. Several factors contributing to the failure include: 1. the strength of Pt wire was decreased due to Si corrosion, which reduces the Pt cross section area; 2. significant grain growth; and 3. high temperature creep. A mechanism was proposed that involves Si/Al vapor reacting with Pt to form deposits, causing Pt thinning followed by high temperature creep and eventually leading to rapture. The proposed failure mechanism is supported by the test conducted in the lab environment.

INTRODUCTION

Thermocouples made of platinum and platinum-rhodium alloys are widely used in high temperature corrosive environments because of their accuracy and resistance to chemical attack and oxidation. However, the life of a thermocouple is largely reduced if metal vapors or other volatiles are in contact with the wires, especially elements like lead, zinc, phosphorous, arsenic, and silicon[1]. This is especially problematic when the use environment is a non-oxidizing or reducing atmosphere. For application where contaminants are present, it is necessary to provide adequate protection, such as using a sealed ceramic or refractory metal tube.

It has been reported that when heated in hydrogen with siliceous refractory, platinum and platinum-rhodium thermocouples become embrittled and easily broken; this occurs due to the reduction of silica in these conditions and subsequent diffusion of Si into the platinum forming platinum silicides (such as Pt Si or Pt_5Si_2) which segregates at the grain boundaries [2]. It is also reported that embrittlement can occur through recrystallization and grain growth for iridium iridium-rhodium thermocouples when heated to 1400°C for a period of time in any atmosphere [3]. Certain semiconductor raw-material process industries require a manufacturing environment that is non-oxidizing and generates metal contaminants that severely impacts the life of the temperature sensor. The premature failure of the sensor causes lower quality of the manufactured products and significantly increases the maintenance cost of the operation. The aim of the present study is to investigate the failure mechanism of a platinum and platinum/rhodium thermocouple at temperatures between 1500-1600 °C in a complex environment with various metal vapors as the first step toward solving premature sensor failure.

EXPERIMENTAL

Platinum and platinum-rhodium (13%) R-type thermocouples made by Watlow were used in the present study. A lab environment was created to simulate the field application environment. In the lab environment, the thermocouple was insulated with an Al_2O_3 tube and placed in Si and Ar atmosphere at 1550°C for a period of time up to 500 h. For control purposes, one of the thermocouples was placed outside the Si atmosphere, but experienced a similar temperature history. The junction of the thermocouple after corrosion was examined by an SEM (scanning electron microscope) to investigate the surface morphology and an EDS (energy dispersive spectroscopy) was used to analyze the chemical composition of the corrosion products.

RESULTS AND DISCUSSION

After being exposed to the Si atmosphere for a period of time at temperature, the thermocouple failed to send the voltage signal. Examination of the thermocouple indicates that wire breakage is responsible for the failure. Further investigation shows that most of the failure is because the Pt wire broke. Silicon- and aluminum-containing deposits were found on Pt and Pt/Rh wire after being exposed to the Si vapor with a variety of deposition morphologies. Fig. 1 shows the elongated shape of the deposit on the Pt wire, as the thermocouple broke after being exposed to the Si atmosphere for 430 h. The diameter of the Pt wire outside the Al_2O_3 insulator tube decreased significantly, while little or no change was recorded for the Pt wire inside the insulator about 0.5 mm below surface. The dark-colored deposit on the Pt wire mainly consists of Si and Al in oxide form. Some of the deposit fell off from the surface, as indicated by the exposed Pt surface (white color in portion in Fig.1 B), which is subject to reacting with Si vapor again later.

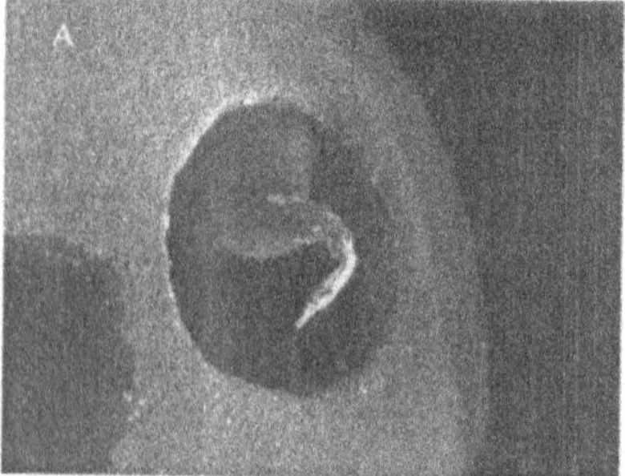

Fig. 1. Morphology of broken Pt wire after 430 h exposure.

Fig 2 shows a dark-colored layer deposit on the Pt wire, which has a composition of Si-Al-Mg in oxide form. Most of the layer at the wire's end flaked off and the Pt surface (lighter in color) was exposed. Significant grain growth of Pt is observed, as indicated by the grains larger than 200 μm.

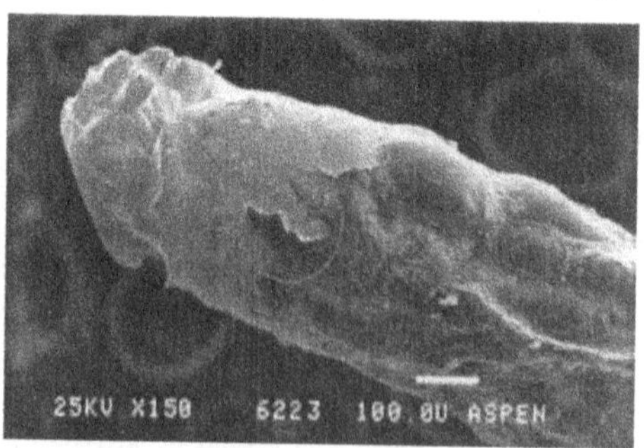

Fig. 2. layered deposit (red circle) flaked off the Pt wire surface.

Fig 3 shows the semi-spherical shape of Si/Al oxide-containing deposits (marked A) on the Pt wire surface marked B. In addition to Si and Al, small amounts of Mg were found in the deposit. The source of the Mg may be from the impurity in the Al_2O_3 tube or due to alien particle contamination. The semispherical shape of the deposit may indicate the temperature of the use environment is higher than the melting temperature of the SiO2-Al_2O_3-MgO composition near the region of cordierite, whose melting temperature is around 1465°C [4].

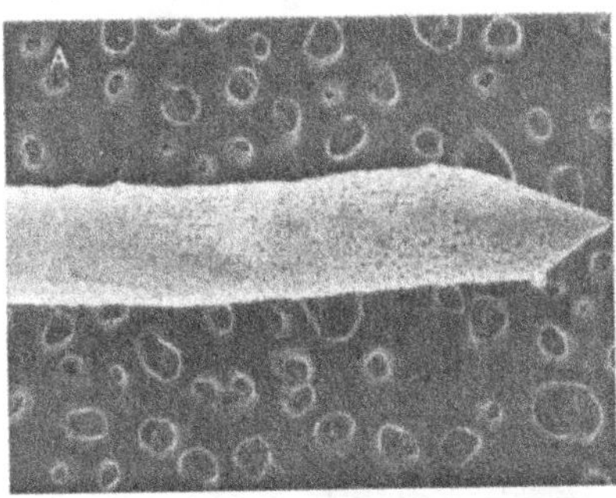

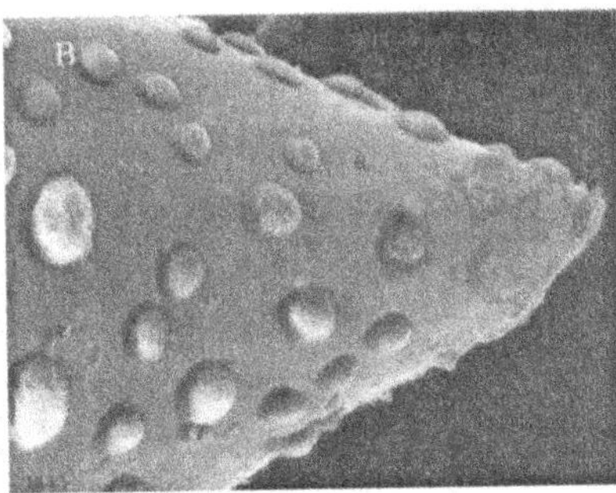

Fig. 3. Semi-spherical deposits on Pt wire surface.

Fig 4A shows a differential corrosion occurring near the grain boundary and grain regions. The concaved shape near the grain boundary is a result of uneven material loss due to corrosion, which indicates a faster corrosion rate as compared to other areas. Many scattered circular-shaped pits were found on the Pt wire. This differential grain boundary corrosion and corrosion pits were also observed on the Pt-Rh wire, as shown in Fig 4B. The removal of the Pt through pitting and grain boundary corrosion reduces the strength of the Pt wire, which assists the tensile failure process.

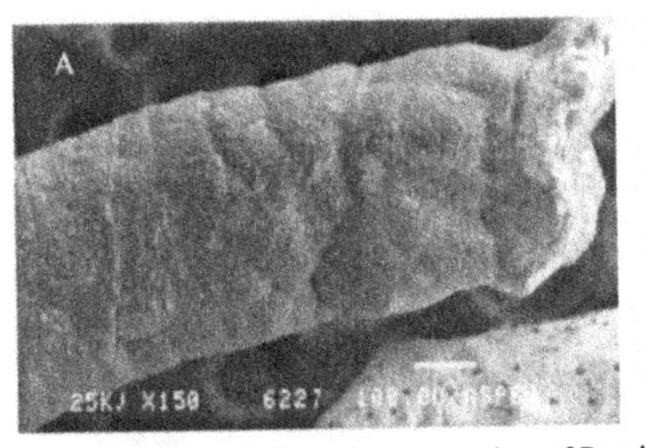

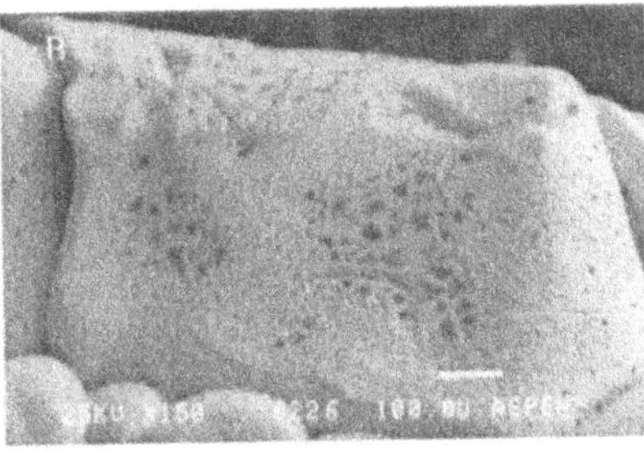

Fig. 4 Pitting and grain boundary corrosion of Pt wire.

The control sample broke after being treated at a similar temperature and for a similar time but without direct exposure to Si vapor. Fig 5 shows the necking of the Pt wire at the breakage. This indicates high temperature creep process involving gravitational force may have occurred, leading to tensile failure. A few spherical deposits containing Si-Al-Ca in oxide form were identified on the surface of wire, although the number of deposits is only a few. Since the Si vapor was confined within the Al_2O_3 tube, it is suggested that the source of the deposit constituents is likely the impurity of the Al_2O_3 tube that holds the sample. Because of the study's small sample size, it is insignificant to compare the life difference between the control and test sample. Further testing will be needed where a statistically significant number of tests will allow clarification of the acceleration factor of silicide formation.

Based on the failure mode and chemistry of deposits on the Pt wire surface, the possible mechanism of Pt wire failure is proposed as the following:

1. Si (vapor) contacts the Pt surface and diffuses into Pt, forming low-temperature brittle platinum silicide on the Pt surface and at grain boundary;
2. Si (vapor) + Al (vapor) + Mg (vapor) + O_2(trace) -> Si/Al/Mg oxides nucleate on platinum silicide and grow a layer or spherical deposit on the Pt surface;
3. Si/Al/Mg oxides flake off the Pt surface due to a mismatch in the thermal expansion coefficient and remove the Pt;
4. The Pt wire becomes thinner after Pt + Al/Si oxides flake off from the surface, which reduces the cross section of the Pt wire and accelerates the high temperature creep that forms the necking;
5. The Pt wire breaks when tensile stress by gravity exceeds the strength of the Pt wire due to cross section reduction.

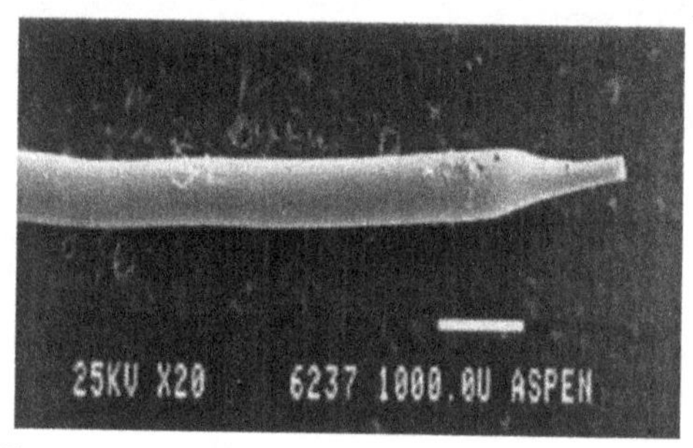

Fig. 5. Control thermocouple shows necking after failure.

CONCLUSION

The primary failure mode of an R-type thermocouple in the Si vapor-rich environment is Pt wire breakage. The key factors contributing to the failure were decreased strength in the Pt wire due to Si corrosion and significant grain growth as well as high temperature creep. The proposed failure mechanism was supported by tests conducted in the lab environment.

REFERENCE

1. J.A. Stevenson, The Metal-clad Thermocouple, Platinum Metals Rev, 4(4), 127-129. (1960)
2. Liu Qingbin, Research on Embrittle Failure of Platinum : 10% Rhodiun-Platinum Thermocouple in Reduction Atmosphere, J. Precious Metals; 2, (2000)
3. H.E. Bennett, The Contamination of Platinum Metal Thermocouples, Platinum Metals Rev., 5(4), 132-133. (1961)
4. Phase diagram for Ceramist, 1964. The American Ceramic Society.

Materials for Extreme Environments

PROTECTION AGAINST OXYDATION, BY CVD OR SPS COATINGS OF HAFNIUM CARBIDE AND SILICON CARBIDE, ON CARBON/CARBON COMPOSITES.

A. Allemand[1,2], Y. Le Petitcorps[1], O. Szwedek[1,2], J.F. Epherre[3], N. Teneze[2], P. David[2]

[1]Université
BORDEAUX
LCTS
33000PESSAC-France

[2]CEA
DAM, Le Ripault,
DMAT
37260MONTS-France

[3]CEA
DAM, CESTA
DSGA
33114LE BARP-France

ABSTRACT

The hafnium carbide compound is an ultra high refractory ceramic; as a result it could be of interest for the protection of carbon/carbon composites against oxidation at high temperatures. However HfC and most of metallic carbides present a non stoechiometric composition with carbon vacancies. As a consequence, the oxidation resistance is poor at low temperatures (500-1000°C). In order to overcome this main drawback the HfC can be associated with silicon carbide (SiC) presenting a better oxidation resistance at lower temperatures.

Two coating routes have been studied; the first one is the Chemical Vapour Deposition which enables to obtain very thin coatings and the second one is the Spark Plasma Sintering technique which permits to get new microstructures of coatings.

On first hand, this study describes the CVD conditions for the deposition of HfC from the metallic hafnium pellets to get hafnium chlorides followed by the reduction of the chlorides by H_2 and the deposition of HfC with the methane as carbon precursor.

On another hand, SPS has permitted to sinter, on carbon substrate, ultra high refractory ceramic powders with a significant amount of SiC. The sintering conditions to obtain an uncracked coating will be presented as well as microstructures and oxidation tests.

INTRODUCTION

Hafnium carbide is one of the most refractory compound (T_f = 3900°C), stiff and tough. HfC could be of interest to protect C/C composites in severe oxidation or ablation conditions [1]. However HfC such as most part of metallic carbides has a very poor oxidation resistance [2], its oxidation starts at 450°C. In order to overcome this main drawback HfC could be associated with a carbide (ex: SiC) presenting a better oxidation resistance [1,3]. HfC and SiC coatings, dense and chemically pure, are obtained by Chemical Vapour Deposition (CVD) technique [1,3].

In this present study two CVD techniques have been used to get two different types of coatings. Low pressure CVD has been chosen to get a HfC and SiC alternate multilayer coating on a C/C composite. In another hand Fluidized Bed CVD (FBCVD) gives an individual SiC coating on each

HfC powder particle. This coated powder has been sintered on C/C substrate by Spark Plasma Sintering.

Thermal behaviour under air of those two types of coatings has been investigated using an arc image furnace.

EXPERIMENTAL

i) Multilayer coatings on C/C substrates by low pressure CVD.

The experimental device is composed of two kilns. The first one (figure 1) is the chlorination device in which the reaction of metallic Hf with HCl at 700°C occurs, which enables to produce the hafnium chloride ($HfCl_4$). This chloride is carried away to the CVD kiln, shown on figure 2, in a pipe heated at 500°C in order to avoid any condensation. A high temperature gas gate can be closed in order to isolate the chlorination device from the CVD kiln. The C/C substrates are placed in this CVD kiln. HfC precursors are hafnium chloride $HfCl_4$ and methane CH_4.

This device enables to get, in one step, a multilayered coating $(SiC/HfC)_n$. To switch from HfC to SiC the high temperature gas gate is closed between the chlorination device and the CVD kiln. At that point, MTS (Méthyltrichlorosilane) carried away by H_2 is the SiC precursor.
Working temperature is 1000°C. For each layer two hours are necessary to obtain a suitable and adherent coating. Low pressure CVD (5 kPa) enables to get a good infiltration of the first SiC layer which leads to a good mechanical behaviour of the coating.

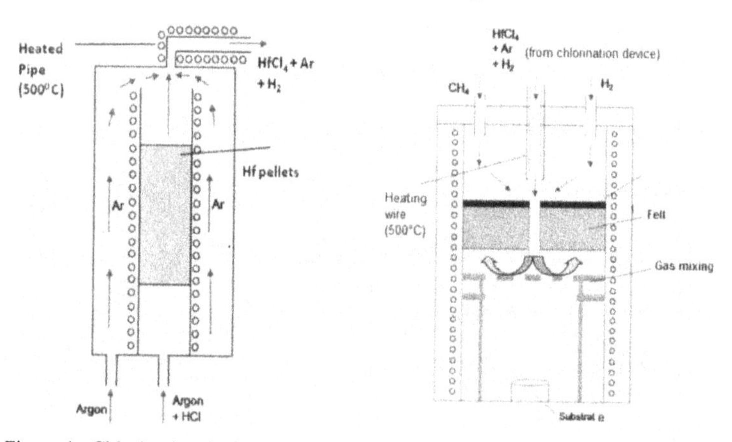

Figure 1 : Chlorination device

Figure 2 : CVD kiln

The C/C samples are disks of 13 mm in diameter and 5 mm in thickness. The C/C is a dense 3D (d > 1.85). Before the CVD treatment the samples are for one part cleaned in an ultrasonic bath of ethanol during 30 minutes then dried and for the other part dipped in a metallic particles slip. Those particles are expected to catalyst the first layer of SiC in order to get SiC whiskers as first layer [4].

ii) C/C protected by sintered coated powder.

In the case of HfC powder coated by SiC, starting powder exhibits a d_{50}=35μm, the SiC coating is achieved by fluidized bed CVD. Hexamethyldisilane is used as a precursor to enhance the SiC coating on HfC powder. The experiment is run under H_2 (transport and reduction) and under N_2 (fluidization gas) at 900°C. It takes 12 hours to get suitable coatings. Thickness of SiC coating is about 1 μm.

Then, this coated powder is directly sintered on C/C substrates by Spark Plasma Sintering device. Sintering conditions are: temperature 1950°C, pressure 75 MPa and dwell time of 5 minutes. The machine is a Dr Sinter 2080 from Syntex (Japan).

Those two coating concepts have been tested under air at temperatures until 2050°C in an arc image furnace at the CEA CESTA. Cycle, in terms of heat flow, is the same for all the tests and enables to reach temperatures, measured by thermocouples, of 1600°C after 50 seconds and 2050°C after 200 seconds.

RESULTS AND DISCUSSION

i) Multilayer coatings on C/C substrates by low pressure CVD.

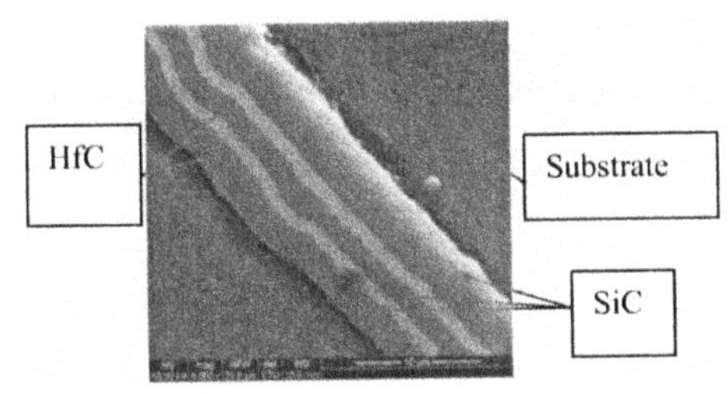

Figure 3: SEM micrography of an alternate multilayer (5 layers) SiC/HfC coating on C/C substrate obtain by CVD.

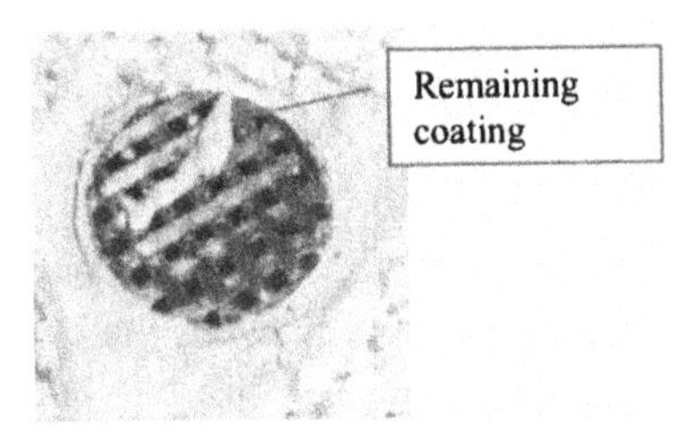

Figure 4: Pictures of a C/C coated with 10 layers coating after a cycle of 205 s, maximum temperature 2050°C.

Figure 3 shows that no cracks inside or between the layers can be seen. SiC/HfC layers are distinct and alternating and they exhibit a thickness of 2 μm. This individual thickness has been determined in order to avoid debonding. This was justified by the lack of reactivity between C/C and SiC and between SiC and HfC. The coating is fully dense and follows the C/C surface. It's even infiltrated inside the open porosity of the C/C substrate.

Oxidation tests have shown that a sample made of a C/C coated with 10 layers (5 SiC, 5 HfC) total thickness of 20 μm, begins to be destroyed after 130 seconds which corresponds to a temperature of 1800°C. The sample is totally destroyed after 160 seconds which corresponds to a temperature of 1900°C. Figure 4 shows the sample after 205 seconds which corresponds to a temperature of 2050°C, one can be seen that there is almost no coating left, the C/C sample appears unprotected. In order to get a coating which can be used during several seconds at 2000°C, a special treatment which brings to a new microstructure has been realised [5].

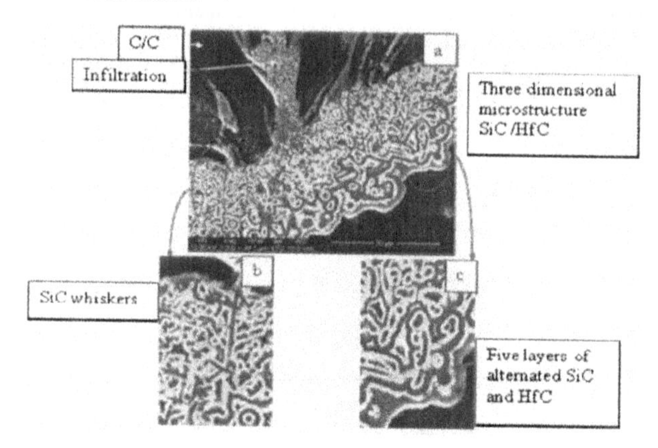

Figure 5: SEM micrography of an alternate multilayer (5 layers) SiC/HfC coating on C/C substrate (with metallic catalyst) obtain by CVD.

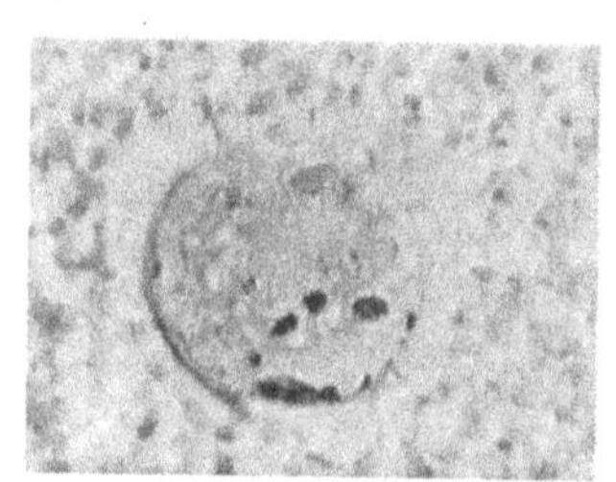

Figure 6: Pictures of a C/C coated with 10 layers coating after a cycle of 205 s, maximum temperature 2050°C.

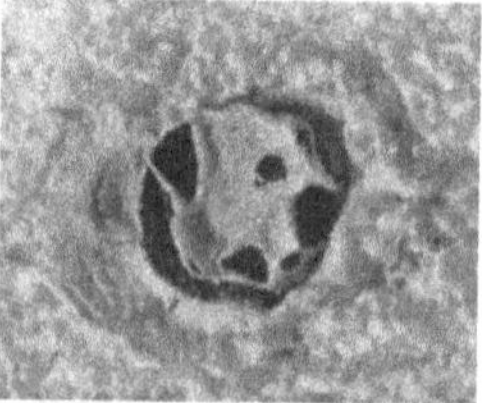

Figure 7: Pictures of a C/C coated with 10 layers coating after a cycle of 370 s, maximum temperature 1900°C.

Figure 5 shows the microstructure of a five layers coating obtained with a sample that have received a layer of catalyst metal particles before the CVD treatment. In such a microstructure the first layer of SiC grows as SiC whiskers of a few microns of diameter. It looks like a "forest of whiskers" over which the second layer is deposed. This second layer composed of HfC matches perfectly the SiC whiskers. Then the third layer composed of SiC covers the previous ones and so on to a total of five layers. This microstructure can be considered as a three dimensional microstructure compared to the one shown in figure 3 in which the layers are stacked on each other which can be considered as a two dimensional microstructure. As the microstructure obtained whithout catalyst this one is fully dense with no cracks inside or between the layers and the coating is still infiltrated inside the open porosities of the C/C sample.

In order to investigate the oxidation behaviour a ten layers coating has been realised. Figure 6 shows the sample after the oxidation test total time of 205 seconds which corresponds to a temperature of 2050°C. The coating is still efficient but the first defects appear and the C/C can be seen underneath the coating. This behaviour must be compared with the one of the two dimensional coating shown in figure 4: for exactly the same oxidation test parameters the three dimensional coating keeps on protecting the C/C at 2050°C whereas the two dimensional coating has been totally destroyed. That can be due to an improvement of the bounding of the three dimensional coating because of the SiC whiskers that act like Velcro® strips. An oxidation test has been run on the three dimensional coating in which the test temperature has been kept constant at 1900°C. This temperature corresponds to the destruction of the two dimensional coating which occurs for a total test time of 160 s. In the case of the three dimensional coating, the figure 7 shows that this one is destroyed after a total test time of 370 seconds. As a conclusion of the oxidation tests we can say that the three dimensional coating is efficient until 2050 °C and at 1900°C the total time of use is multiplied by two with respect to the two dimensional coating.

ii) C/C protected by sintered coated powder

Figure 8 shows that, after sintering, the coating is dense (95%TD), no cracks can be seen and its thickness is about 300µm. At a higher magnification SiC can be seen at grain boundaries around HfC powder.

In terms of oxidation test on coated powder a TGA until 1500°C under air has been run. Total TGA time is 2h and 30 min. This one is presented on figure 9 where mass gain (%) is presented versus temperature (°C). Oxidation of non coated HfC powder begins at 380°C and is totally finished at 800°C. In reverse SiC is rather stable even at high temperature. The most interesting thing is the fact that the HfC powder coated with 1µm of SiC almost follows the behaviour of pure SiC.

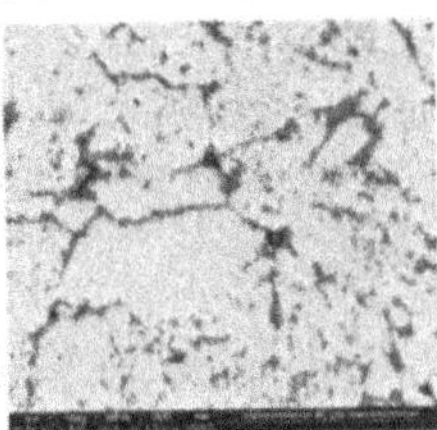

Figure 8: SEM micrographs of a coating made of a sintered (SPS) HfC powder coated with SiC over a C/C substrate.

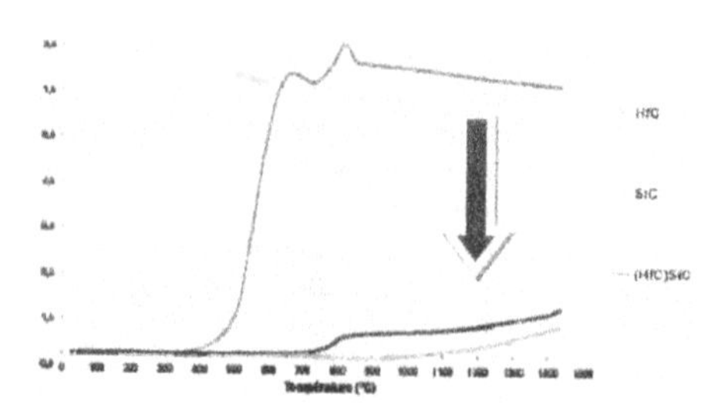

Figure 9: TGA under air until 1500°C of pure HfC, HfC coated with 1μm SiC and pure SiC.

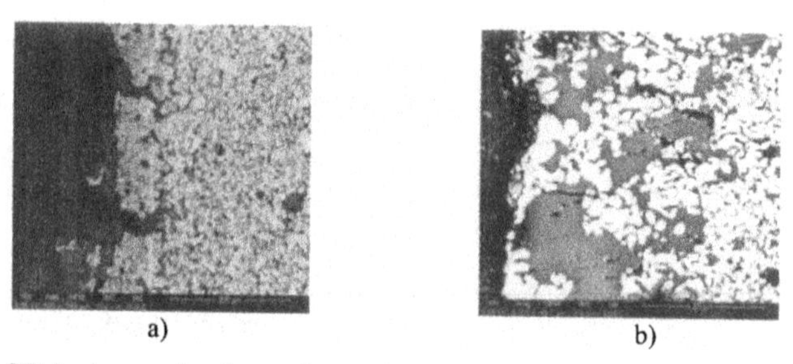

a) b)

Figure 10: SEM micrographs of a coating made of a SPS sintered HfC powder coated with SiC after oxidation at 2050°C b) BSE.

Figure 10 presents the microstructure of the coating after the oxidation test. This test has been run in an arc image furnace. Maximum temperature reached is 2050°C and time of exposure is 205 seconds [6]. In terms of microstructure one can be seen that the oxidation occurs until 100μm deep. Under this oxidised layer, the coating is safe without any cracks. At higher magnification, back scattering electrons show, in the oxidation zone, two chemical compositions and a microstructure pointing out a melted zone. An EDS analysis has shown that the darker zone is composed of Si, Hf and O. We can think that it's probably $HfSiO_4$. The light zone is composed of Hf and O. We can think of HfO_2. Further investigations have to be run.

About fusion temperatures $HfSiO_4$ melts at 1750°C and HfO_2 at 2800°C. So, with respect to the previous observations and to the fusion temperatures, in terms of oxidation mechanisms, HfC and SiC together, give, during oxidation at temperatures higher than 1800°C, a glass mainly made of $HfSiO_4$ which is fluid enough to glaze the surface of HfO_2 (formed during the oxidation of HfC grain). That oxidation behaviour is explained on figure 11.

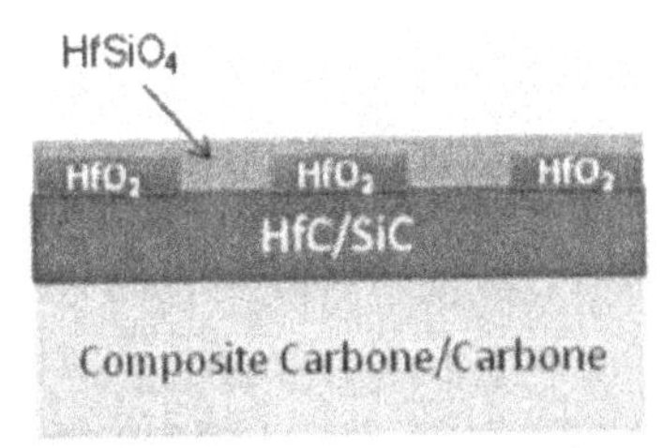

Figure 11: Oxidation behaviour of an HfC/SiC coating over a C/C at temperatures higher than 1800°C.

CONCLUSION

Various coatings morphologies synthesized by two CVD techniques have been tested up to 2000°C under air. The two concepts studied have been patented [5, 6]. On one hand the low pressure CVD enables to get thin layers and a final coating of several dozen microns, on the other hand the sintering of a coated powder by fluidized bed CVD enables to get thicker coatings of hundreds microns. Each kind of coatings is dense with no cracks and they all bind the surface of the C/C substrate. Those two kinds of coating able to protect during several hundred seconds a C/C at a temperature as high as 2000°C.

ACKNOWLEDGMENT

The authors want to greatly thank Mr Nisio from MPA Industry for his help in all the steps of chlorination and low pressure CVD devices, the CEA CESTA for experiments on the arc image furnace, the "Plateforme Nationale de frittage flash" PNF² in Toulouse for SPS sintering, Mr V.Frotté from CEA RIPAULT for the TGA and LIFCO Industry for the powder coating development.

REFERENCES

[1] B. Bavarian and V. Arrieta, Proceedings of the 35th International Society of Advanced Materials Process Engineers Symposium, **35**, 1348 (1990).
[2] W.J. Lackey, J.A. Hanigofsky, A. John, G.B. Freeman and B. Garth, J. Am. Ceram. Soc., **73**, 1593-1598 (1990).
[3] R.B.Kaplan and al, US Pat. 5283109 (1994)
[4] W.J. Kim and al. "Effect of a SiC whisker formation on the densification of Tyranno SA/SiC composites fabricated by the CVI process" in Fusion Engineering and Design, **81**, 931-936 (2006).
[5] A. Allemand, Y. Le Petitcorps, O. Szwedek, J.F. Epherre, French Patent application N°1160815 (2011).
[6] A. Allemand, Y. Le Petitcorps, O. Szwedek, L. Bianchi, French Patent application N°1058711 (2010).

PROPERTIES OF HOT-PRESSED Ti_2AlN OBTAINED BY SHS PROCESS

L. Chlubny, J. Lis, M.M. Bućko, D. Kata
AGH - University of Science and Technology, Faculty of Materials Science and Ceramics, Department of Technology of Ceramics and Refractories, Al. Mickiewicza 30, 30-059, Cracow, Poland

ABSTRACT

Some of ternary materials in the Ti-Al-N system, called MAX-phases, are characterised by heterodesmic layer structure. Their specific structure consisting of covalent and metallic chemical bonds influence their semi-ductile features, locating them on the boundary between metals and ceramics. These features may lead to many potential applications, for example as a part of ceramic armour. Ti_2AlN is one of these nanolaminate materials. Self-propagating High-temperature Synthesis (SHS) was applied to obtain sinterable powders of Ti_2AlN precursors. Intermetallic compounds were used as precursors in the synthesis. For densification of obtained powders hot-pressing technique was used. Phase compositions of dense samples were examined by XRD method. Obtained samples contained over 75 wt.% of MAX phase, other phases were Ti_3Al, Ti_3AlN and TiN. Properties, such as hardness, bending strength, fracture toughness and elastic properties were determined.

INTRODUCTION

Among many advanced ceramic materials such as carbides or nitrides there is a group of ternary compounds referred in literature as H-phases (or Hägg-phases), Nowotny-phases, thermodynamically stable nanolaminates and at last but not least - MAX phases. These compounds have a $M_{n+1}AX_n$ stoichiometry, where M is an early transition metal, A is an element of A groups (mostly IIIA or IVA) and X is carbon and/or nitrogen. Heterodesmic structures of these phases are hexagonal, P63/mmc, and specifically layered. They consist of alternate near close-packed layers of M_6X octahedrons with strong covalent bonds and layers of A atoms located at the centre of trigonal prisms. The M_6X octahedral, similar to those forming respective binary carbides, are connected one to another by shared edges. Variability of chemical composition of the nanolaminates is usually labeled by the symbol describing their stoichiometry, e.g. Ti_2AlC represents 211 type phase and Ti_3AlC_2 – 312 type. Structurally, differences between the respective phases consist of the number of M layers separating the A-layers: in the 211's there are two, whereas in the 321's three M-layers. This new group of compounds was described in details in work of Nowotny et.al, Barsoum and Z. Lin et al. [1-3, 5]. The layered, heterodesmic structure of MAX phases led to an extraordinary set of properties. These materials combine properties of ceramics like high stiffness, moderately low thermal expansion coefficient and excellent thermal and chemical resistance with low hardness, good compressive strength, high fracture toughness, ductile behaviour, good electrical and thermal conductivity characteristic for metals. They can be used to produce ceramic armour based on functionally graded materials (FGM) or as a matrix in ceramic-based composites reinforced by covalent phases.

The SHS is a method that allows obtaining lots of covalent materials such as carbides, borides, nitrides, oxides and intermetallic compounds. At the base of this method lays utilization of exothermal effect of chemical synthesis, which can proceed in powder bed of solid substrates or as a filtration combustion. An external source of heat is used to initiate the process and then the self-sustaining reaction front is propagating through the bed of substrates. This process could be initiated by local ignition or by thermal explosion. The form of synthesized material depends on kind of a precursor used for the synthesis and technique that was applied. Typical feature of this reaction are low energy consumption, high temperatures obtained during the process, high efficiency and simple apparatus. The lack of control of the process is the disadvantage of this method[4].

The objective of this work was obtaining of dense samples of Ti_2AlN materials manufactured from powders synthesized by SHS method and sintered by hot-pressing method and examining of some of their mechanical properties.

PREPARATION

Following the experience gained while synthesizing ternary materials such as Ti_3AlC_2 and Ti_2AlC [7, 8, 12], as well as earlier experiments with Ti_2AlN synthesis [11, 13, 14], intermetallic materials in the Ti-Al system were used as a precursors for synthesis of Ti_2AlN powders.

Due to relatively low availability of commercial powders of intermetallic materials in the Ti-Al system, it was decided to synthesize them by SHS method. At the first stage of the experiment TiAl powder was synthesized by SHS method [7]. Titanium hydride powder, TiH_2, and metallic aluminium powder were used as sources of titanium and aluminium. The mixture for SHS reaction had a molar ratio of 1:1 (equations 1).

$$TiH_2 + Al \rightarrow TiAl + H_2 \quad (1)$$

Powder obtained in the SHS synthesis was ground in rotary-vibratory for 8 hours in isopropanol to the grain size ca. 10 μm[12].

The next stage was synthesis of Ti_2AlN, which was conducted by the SHS method with a local ignition system. The precursors used for a synthesis was SHS derived TiAl powder and commercially available powder of titanium and pure nitrogen. The mixture of precursors for a SHS synthesis was set in appropriate stoichiometric ratio and is presented in equation 2.[13]

$$2\ TiAl + 2\ Ti + N_2 \rightarrow 2\ Ti_2AlN \quad (2)$$

The X-ray diffraction analysis method was applied to determine phase composition of synthesised materials and to confirm their usability for the further processing. The basis of phase analysis were data from ICCD [9]. Phase quantities were determined by Rietveld analysis [10]. The density of obtained powders was determined by use of Micromeritics Accu-Pyc 1330 helium pycnometer. Average grain size was measured with use of Micromeritics ASAP 2010.

In the next step hot-pressing technique was applied for the densification procedure. Powder was sintered in the Thermal Technologies 2000 hot press, at temperature 1300°C, annealing time in maximum temperature was 1 hour, sintering was conducted in constant flow of nitrogen under a pressure 24.1 MPa. Diameter of obtained samples was 2.5 and 7.6 cm and height of samples was 0.5 cm. The possibility of this approach was proven by Z.J. Lin et al, during hot pressing of TiN + Al + Ti mixture up to 1400 °C under a pressure of 25 MPa as well as by previous researches of authors on the Ti-Al-C-N system [6, 11, 14, 15].

Sintered samples were examined to estimate phase quantities of particular dense bodies. Scanning electron microscopy (FEI Europe Company Nova Nano SEM 200) was applied to examine morphology of samples. The relative density was measured by the hydrostatic weighing. Elastic properties were established by ultrasonic measurement with use of UZP-1 INCO-VERITAS defectoscope. Hardness was measured by the Vickers indentation method (FV-700 / Future Tech) and bending strength by the three point bending test (Zwick/Roell BTC-FR2.5TS.D14).

RESULTS AND DISCUSSION

The XRD analysis proved that TiAl synthesised by SHS method was almost phase pure and contained only about 5% of Ti_3Al impurities, as it is presented on Figure 1[11].

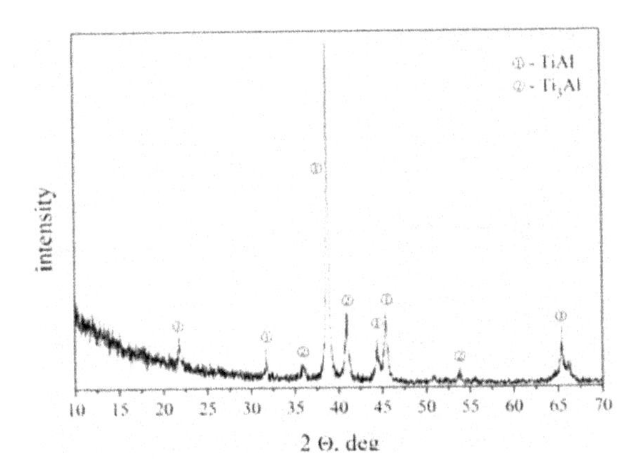

Also phase analysis of product of Ti_2AlN SHS synthesis showed presence of the MAX phase. The obtained powder in this procedure was mixture of five phases. Basing on the experience from previous work, powder was qualified as a precursor for sintering of dense material. The XRD pattern of obtained powder is presented on Figure 2[15]. The phase composition of SHS product is presented in Table I. The parameters of MAX phase powder are presented in Table II.

Table I. Products of SHS synthesis of Ti_2AlN phase composition

Precursor	Composition, wt %.
$TiAl + Ti + N_2$	57% Ti_2AlN, 24% TiN, 11% Ti_3Al, 8% Ti_3AlN [11]

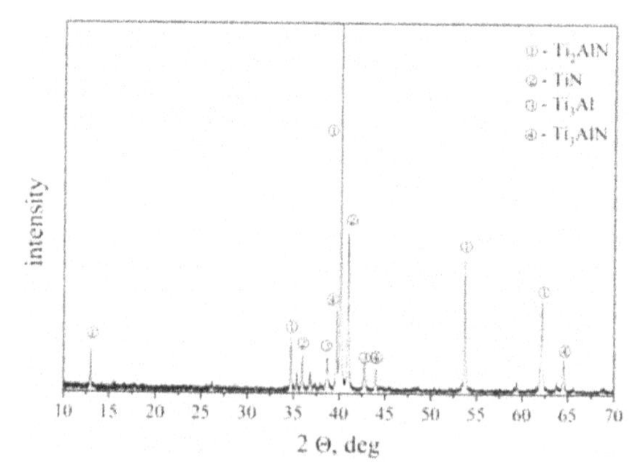

Figure 2. XRD pattern of product of 2 TiAl + 2 Ti + N_2 synthesis[15]

Table II. Parameters of SHS derived Ti_2AlN powder

Parameter	Value
Density	4.187 ± 0.002 g/cm^3.
Average grain size	16.96μm

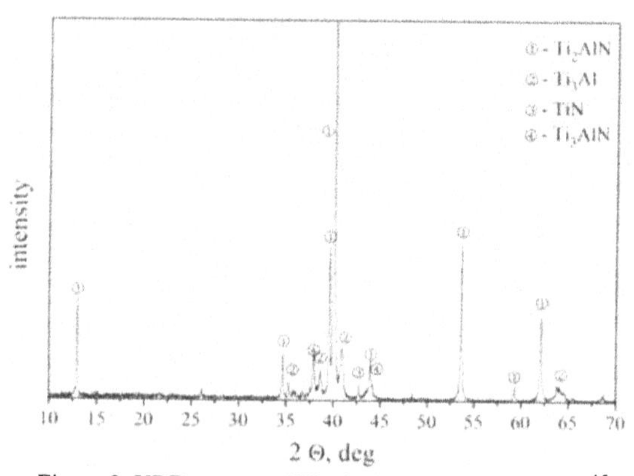

Figure 3. XRD pattern of Ti_2AlN hot pressed at 1300°C[15].

Observation made on hot-pressed sample showed that further chemical reactions are proceeding during the sintering process. It was observed that amount of MAX phase is increasing in the system at cost of other phases. In 1300°C dense, polycrystalline material, containing over 75 wt.% of Ti_2AlN was obtained. The other phases were 13.4 wt.% Ti_3Al, 9.6 wt.% Ti_3AlN and 1.3 wt.% TiN. The XRD result is presented on Figure 3[15]. Above this temperature hexagonal structure changes into regular structure and decomposition of MAX phases takes place. This process is characteristic for a lot of MAX phases, for example it occurs in case of Ti_3SiC_2.

The SEM pictures of sintered materials presents characteristic for MAX phases plate-like grains (Figure 4).

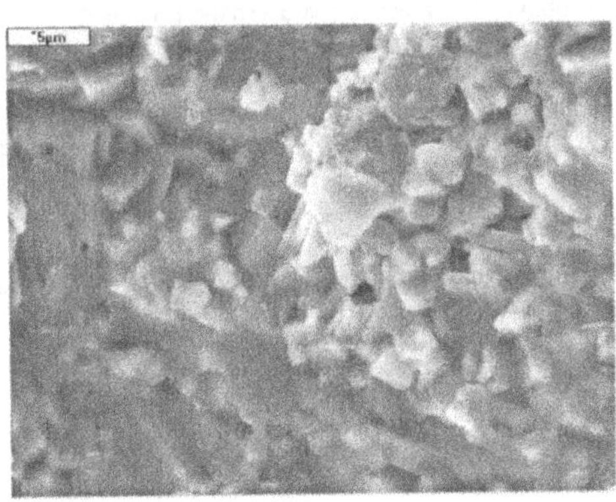

Figure 4. SEM picture of Ti_2AlN hot pressed at 1300 °C

Results of mechanical properties examinations, such as density, elastic properties, Vickers hardness and bending strength of hot pressed Ti_2AlN are presented in Table III. Moreover, during the Vickers hardness test, plastic deformation zones around indentation marks were observed.

Table III. Characteristic of hot-pressed Ti_2AlN material.

Parameter	Value
Relative density	4.15 ± 0.01 g/cm^3
Young's modulus	234 ± 20 GPa
Shear modulus	95 ± 4 GPa
Poisson ratio	0.23 ± 0.03
Vickers hardness	5.8 ± 0,8 GPa
Bending strength	141.8 ± 11.5 MPa

CONCLUSIONS

It was proved that obtaining of sinterable precursors of Ti_2AlN materials by Self-propagating High-temperature Synthesis (SHS) is possible. This method appears to be an effective, efficient and economic. Use of intermetallic compounds, particularly TiAl, is necessary in case of this synthesis.

During densification process by hot-pressing method, reaction sintering is observed. This leads to significant increase of amount of Ti_2AlN in the sample volume. These phenomena were also observed in case of other MAX phases such as Ti_3SiC_2, Ti_3AlC_2 and Ti_2AlC.

Hot pressing of these powders did not allow obtaining single phase Ti_2AlN material, but dense material in which ternary phase is a dominating phase (over 75 wt.%).

Morphology of obtained samples is characteristic for this specific group of materials and consists of plate-like grains, which influence cracking mechanism of these materials. Elastic and mechanical properties are also comparable to other materials from this group. Further researches on fracture toughness and mechanism of strains within the material as well as some of thermal and electrical properties will be conducted. Also researches focused on increasing of content of ternary compound both in SHS deriver powder and in dense material will be conducted.

ACKNOWLEDGMENTS

This work was supported by the Polish Ministry of Science and Higher Education under the grant no. POIG.01.01.02-00-097/09-02 “Nowe materiały konstrukcyjne o podwyższonej przewodności cieplnej”.

REFERENCES

[1] W. Jeitschko, H. Nowotny, F.Benesovsky, Kohlenstoffhaltige ternare Verbindungen (H-Phase). *Monatsh. Chem.* **94**, 1963, p 672-678

[2] H. Nowotny, Structurchemie Einiger Verbindungen der Ubergangsmetalle mit den Elementen C, Si, Ge, Sn. *Prog. Solid State Chem.* **2** 1970, p 27

[3] M.W. Barsoum: The MN+1AXN Phases a New Class of Solids; Thermodynamically Stable Nanolaminates- *Prog Solid St. Chem.* **28**, 2000, p 201-281

[4] J.Lis: Spiekalne proszki związków kowalencyjnych otrzymywane metodą Samorozwijającej się Syntezy Wysokotemperaturowej (SHS) - *Ceramics 44* : (1994) (*in Polish*)

[5] Z. Lin, M. Li, Y. Zhou: TEM Investigations on Layered Ternary Ceramics; *J Mater Sci Technol* **23** (2007) 145-165

[6] Z.J. Lin, M.J. Zhuo, M.S. Li, J.Y. Wanga, Y.C. Zhou: Synthesis and microstructure of layered-ternary Ti_2AlN ceramic; Scripta Mater **56** (2007) 1115–1118

[7] L. Chlubny, M.M. Bucko, J. Lis “Intermetalics as a precursors in SHS synthesis of the materials in Ti-Al-C-N system” *Advances in Science and Technology*, **45**, 2006, p 1047-1051

[8] L. Chlubny, M.M. Bucko, J. Lis “Phase Evolution and Properties of Ti_2AlN Based Materials, Obtained by SHS Method” Mechanical Properties and Processing of Ceramic Binary, Ternary

and Composite Systems, *Ceramic Engineering and Science Proceedings*, Volume **29**, Issue 2, 2008, Jonathan Salem, Greg Hilmas, and William Fahrenholtz, editors; Tatsuki Ohji and Andrew Wereszczak, volume editors, 2008, p 13-20

[9] "Joint Commitee for Powder Diffraction Standards: International Center for Diffraction Data"

[10] H. M. Rietveld: "A profile refinement method for nuclear and magnetic structures." J. Appl. Cryst. **2** (1969) p. 65-71

[11] L. Chlubny: New materials in Ti-Al-C-N system. - PhD Thesis. AGH-University of Science and Technology, Kraków 2006. (*in Polish*)

[12] L. Chlubny, J. Lis, M.M. Bucko: Preparation of Ti_3AlC_2 and Ti_2AlC powders by SHS method MS&T Pittsburgh 09: Material Science and Technology 2009, 2009, p 2205-2213

[13] L. Chlubny, J. Lis, M.M. Bucko "Titanium and Aluminium Based Compounds as a Precursors for SHS of Ti_2AlN" Nanolaminated Ternary Carbides and Nitrides, *Ceramic Engineering and Science Proceedings*, Volume **31**, Issue 10, 2010, Sanjay Mathur and Tatsuki Ohji, volume editors, 2010, p 153-160

[14] L. Chlubny, J. Lis, M. M. Bucko "Influence of Sintering Conditions on Phase Evolution of SHS-Derived Materials in the Ti-Al-N System" ECERS 2007, *Proceedings of the 10th international conference of the European Ceramic Society* : June 17–21, 2007 Berlin. eds. J. G. Heinrich, C. G. Aneziris. Baden-Baden, Göller Verlag GmbH, 2007, p. 1155–1158.

[15] L. Chlubny, J. Lis, M. M. Bucko, D.Kata "Pressureless Sintering and Hot-Pressing of Ti_2AlN Powders Obtained Powders Obtained by SHS Process", *Proceedings of the 35th International Conference on Advanced Ceramic and Composites, January 23-28, 2011 Daytona Beach, Florida – Advanced Ceramic Coatings and Interfaces VI*, Volume **32**, Issue 3, volume editors Sujanto Widjaja and Dileep Singh, 2011, p 161-168

A COMPARATIVE STUDY OF DECOMPOSITION KINETICS IN MAX PHASES AT ELEVATED TEMPERATURE

I.M. Low[1], W.K. Pang[1,2]
[1]Department of Imaging & Applied Physics, Curtin University of Technology, GPO Box U1987, Perth, WA 6845.
[2]Department of Materials Science & Engineering, Tatung University, Taipei, Taiwan.

ABSTRACT

The role of pore microstructures on the susceptibility of MAX phases (Ti_3SiC_2, Ti_3AlC_2, Ti_2AlC, Ti_2AlN_2, Ti_4AlN_3) to thermal dissociation at 1300-1550 °C in high vacuum has been studied using *in-situ* neutron diffraction. Above 1400 °C, MAX phases decomposed to binary carbide (e.g. TiC_x) or binary nitride (e.g. TiN_x), primarily through the sublimation of A-elements such as Al or Si, forming in a porous surface layer of MX_x. Positive activation energies were determined for decomposed MAX phases with coarse pores but a negative activation energy when the pore size was less than 1.0 μm. The role of pore microstructures on the decomposition kinetics is discussed.

INTRODUCTION

MAX phases exhibit a unique combination of characteristics of both ceramics and metals with unusual mechanical, electrical and thermal properties [1-6]. These materials are nano-layered ceramics with the general formula $M_{n+1}AX_n$ (n = 1–3), where M is an early transition metal, A is a group A element, and X is either carbon and/or nitrogen. Similar to ceramics, they possess low density, low thermal expansion coefficient, high modulus and high strength, and good high-temperature oxidation resistance. Like metals, they are good electrical and thermal conductors, readily machinable, tolerant to damage, and resistant to thermal shock. The unique combination of these interesting properties enables these ceramics to be promising candidate materials for use in diverse fields which include automobile engine components, heating elements, rocket engine nozzles, aircraft brakes, racing car brake pads and low-density armour.

However, the high-temperature thermochemical stability in MAX phases has hitherto generated much controversy among researchers. For instance, several researchers have reported that Ti_3SiC_2 became unstable at temperatures greater than 1400°C in an inert atmosphere (e.g. vacuum, argon or nitrogen), by dissociating into Si, TiC_x and/or $Ti_5Si_3C_x$ [7-12]. A similar phenomenon has also been observed for Ti_3AlC_2 whereby it decomposes in vacuum to form TiC and Ti_2AlC [13-16].

In other studies, Zhang *et al.* [17] reported Ti_3SiC_2 to be thermally stable up to 1300°C in nitrogen, but above this temperature drastic degradation and damage occurred due to surface decomposition. Feng *et al.* [18] annealed the Ti_3SiC_2-based bulk samples at 1600 °C for 2h and 2000 °C for 0.5 h in vacuum (10^{-2} Pa) and found that TiC_x was the only phase remaining on the surface. According to Gao *et al.* [19] the propensity of decomposition of Ti_3SiC_2 to TiC_x was related to the vapour pressure of Si, i.e., the atmosphere where the Ti_3SiC_2 exits. They believed that the partial pressure of Si plays an important role in maintaining the stability of Ti_3SiC_2 whereby it has a high propensity to decompose in N_2, O_2 or CO atmosphere at temperatures above 1400°C. This process of surface-initiated phase decomposition was even observed to commence as low as 1000–1200°C in Ti_3SiC_2 thin films during vacuum annealing [20]. The large difference in observed decomposition temperatures between bulk and thin-film Ti_3SiC_2 has been attributed to the difference in diffusion length scales involved and measurement sensitivity employed in the respective studies. In addition,

Ti_3SiC_2 has also been observed to react readily with molten Al, Cu, Ni and cryolite (Na_3AlF_6) at high temperatures.

In contrast, Barsoum and co-workers [21] have shown that Ti_3SiC_2 was thermodynamically stable up to at least 1600 °C in vacuum for 24h and in argon atmosphere for 4h. They further argued that the reduced temperature at which Ti_3SiC_2 decomposed as observed by others was due to the presence of impurity phases (e.g. Fe or V) in the starting powders which interfered with the reaction synthesis of Ti_3SiC_2, and thus destabilized it following prolonged annealing in an inert environment [22]. However, mixed results have been reported by Radhakrishnan *et al.* [23]. In their investigation, Ti_3SiC_2 was shown to be stable in a tungsten-heated furnace for 10h at 1600°C and 1800°C in an argon atmosphere, but dissociated to TiC_x under the same conditions when using a graphite heater.

These conflicting results suggest that the thermochemical stability of MAX phases is still poorly understood although its susceptibility to thermal decomposition is strongly influenced by factors such as purity of powders and sintered materials, temperature, vapour pressure, atmosphere, and the type of heating elements used. In addition, the nature of microstructure of the decomposed surface layer formed during annealing remains controversial, especially in relation to the role of pore sizes in the decomposition kinetics at the near surface.

In this paper, we have conducted a comparative study to elucidate the role of pore microstructures on the decomposition kinetics of several MAX phases during vacuum annealing in the temperature range 1000-1800°C. The effect of pore-size on the activation energy of decomposition was evaluated using the Arrhenius equation.

EXPERIMENTAL PROCEDURE

Hot-pressed and fully dense cylindrical MAX-phase samples (Ti_3SiC_2, Ti_3AlC_2, Ti_2AlC, Ti_2AlN_2, Ti_4AlN_3) of 15 mm diameter and 25 mm height were used for this study. The samples were hot-pressed in an argon atmosphere at a heating rate of 50°C/min until 1400-1450°C, where it was held for 2 h. The pressure during the hot-pressing was 30 MPa.

In-situ neutron diffraction was used to monitor the structural evolution of the phase decomposition of MAX phases at high-temperature in real time. Diffraction patterns were collected using WOMBAT at OPAL or the Polaris at ISIS, both are high-intensity medium-resolution powder diffractometers. Samples were mounted in a high-temperature niobium furnace which was fitted with a thin tantalum foil element and tantalum and vanadium heat shields that allows it to reach 2000°C, and operates under a high dynamic, i.e., it is continuously evacuated with a vacuum of 10^{-6} – 10^{-8} torr. A precision electronic scale was used to weigh the sample before it was loaded into the furnace.

The sample was held by vanadium wire and heated to 1000°C at a heating rate of 10°C/min and thereafter at 5°C/min to 1550°C. The sample was dwelled for 30 minutes at 1000-1300 and 200 minutes at 1400-1550°C. Diffraction patterns were collected every minute during the experiment. After the last measurement at 1550°C, the furnace was cooled to room temperature, and the sample carefully removed from the furnace and weighed to determine the mass lost through evaporation.

The Rietveld method was used to compute the phase contents as a function of temperature during vacuum-annealing of each MAX phase. The decomposition rates (k) at different temperatures were calculated and the corresponding apparent activation energies (E) were determined using the Arrhenius' equation,

$$k = k_0 \exp\left(-\frac{E}{RT}\right) \qquad (1)$$

where k_0 is constant, R is gas constant (8.3145 $J\cdot K^{-1}\cdot mol^{-1}$) and T is absolute temperature [24].

Scanning electron micrographs of vacuum-annealed or decomposed MAX phases were acquired using the Zeiss EVO 40XVP SEM with an accelerating voltage of 15 keV. The samples were not gold-coated before the microstructure examination and the images were taken using secondary electrons.

RESULTS AND DISCUSSION

Thermal stability and phase transitions

The phase transitions in several *MAX* phases and their relative phase abundances at various temperatures as revealed by in-situ neutron diffraction is shown in Figure 1. A weight loss of ~ 4% was observed for decomposed Ti_3SiC_2 which may be attributed to the release of gaseous Ti and Si by sublimation during the decomposition process. For Ti_3AlC_2, its decomposition into TiC and Ti_2AlC as lower order or intermediate phase was observed at ≥ 1400°C. However, at higher temperatures, when compared to TiC, a smaller growth rate for Ti_2AlC may indicate that Ti_2AlC experienced further decomposition into TiC via the sublimation of Al, similar to decomposition of Ti_3SiC_2. In contrast to Ti_3AlC_2, no intermediate or lower order phase was observed for the decomposition of Ti_3SiC_2. This difference can be attributed to the fact that Ti_3SiC_2 is the only stable ternary phase in Ti-Si-C diagram. Fig. 2(d) shows the excellent stability of Ti_2AlN at 1500°C for up to 350 minutes.

In general, a weight loss of up to 20% was observed as a result of decomposition for all the MAX phases can be attributed to the release of gaseous Al by sublimation during the decomposition process because the vapour pressures of the A elements exceed the ambient pressure of the furnace (i.e. $\leq 5 \times 10^{-5}$ torr) at ≥1500°C. Since the vapor pressure of a substance increases non-linearly with temperature according to the Clausius-Clapeyron relation [25], the volatility of A elements will increase with any incremental increase in temperature.

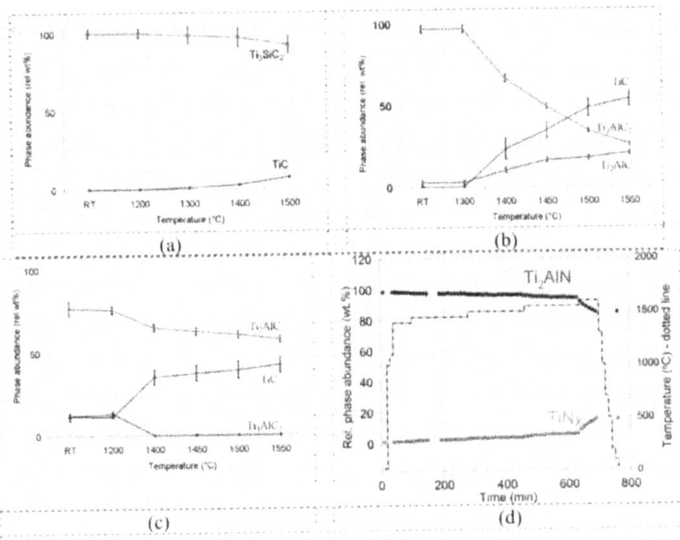

Fig. 1: Phase abundance as a function of temperature for the decomposition of (a) Ti_3SiC_2, (b) Ti_3AlC_2, (c) Ti_2AlC, and (d) Ti_2AlN in vacuum.

It is well known that A elements such as Si and Al have high vapour pressure and become volatile at elevated temperature [26]. Thus, at the temperature of well over 1500°C used in this study, both Al and Si should become volatile and sublime readily and continuously in a dynamic environment of high vacuum. When the vapor pressure becomes sufficient to overcome ambient pressure in the vacuum furnace, bubbles will form inside the bulk of the substance which eventually appears as voids on the surface of decomposed MAX phase. The evidence of surface voids formation can be clearly discerned from the scanning electron micrographs of decomposed MAX phases shown in Fig. 2. Since Si has a lower vapour pressure than Al [26], it helps to explain why Ti_3SiC_2 is more resistant to decomposition than Ti_3AlC_2 or Ti_2AlN. In all cases, the kinetics of decomposition process is driven mainly by a highly restricted out-diffusion and sublimation of high vapour pressure *A* element (e.g. Al, Si) from the bulk to the surface of the sample and into the vacuum, i.e.,

$$M_{n+1}AX_n \rightarrow M_{n+1}X_n + A$$
$$M_{n+1}X_n \rightarrow (n+1)MX_{n/(n+1)} \qquad (2)$$

Role of Pore Microstructures on Decomposition Kinetics

The activation energies calculated from the Arrhenius equation for the five *MAX* phases and the proposed reactions are summarized and listed in Table 1. All the calculated activation energies are positive except for bulk Ti_3AlC_2. However, when powder of Ti_3AlC_2 was used a positive activation energy was obtained (see Figure 3) which implies the importance of pore microstructures in the decomposition kinetics. A negative activation energy indicates that the rate of decomposition in Ti_3AlC_2 decreased with increasing temperature due to the formation of a dense TiC surface layer with very fine pores (<1.0 μm) which exert an increasing resistance to the sublimation process as the temperature increases (Fig. 2d). In contrast, a more porous decomposed layer with coarser pores (>2.0 μm) formed in other MAX phases and in powdered Ti_3AlC_2 which enabled the sublimation of Al or Si to progress with minimum resistance, thus resulting in an increasing rate of decomposition with temperature (Fig. 2(a-c)). In short, the pore sizes play a critical role in determining the value of activation energy and the rate of decomposition. Thus, the ability to manipulate the pore microstructure either through densification to reduce pore-size or engineering of pore-free microstructure will allow the process of decomposition in MAX phases to be minimized or arrested.

Table 1: Comparison of the kinetics of decomposition in five MAX-phase samples.

MAX phase	Activation energy (kJ mol^{-1})	Pore size (μm)	Proposed reactions
Ti_3SiC_2	169.6	1.0 - 3.0	$Ti_3SiC_{2(s)} \rightarrow 3TiC_{0.67(s)} + Si_{(g)}$
Ti_3AlC_2 (bulk)	-71.9	0.5 - 0.8	$Ti_3AlC_{2(s)} \rightarrow 3TiC_{0.67(s)} + Al_{(g)}$
Ti_3AlC_2 (powder)	71.9	>1.0	$Ti_3AlC_{2(s)} \rightarrow 3TiC_{0.67(s)} + Al_{(g)}$
Ti_2AlC	85.7	2.0 – 10.0	$Ti_2AlC_{(s)} \rightarrow 2TiC_{0.5(s)} + Al_{(g)}$
Ti_2AlN	573.8	2.0 – 8.0	$Ti_2AlN_{(s)} \rightarrow 2TiN_{0.5(s)} + Al_{(g)}$
Ti_4AlC_3	410.8	1.8 – 3.0	$Ti_4AlN_3 \rightarrow 4TiN_{0.75} + Al_{(g)}$

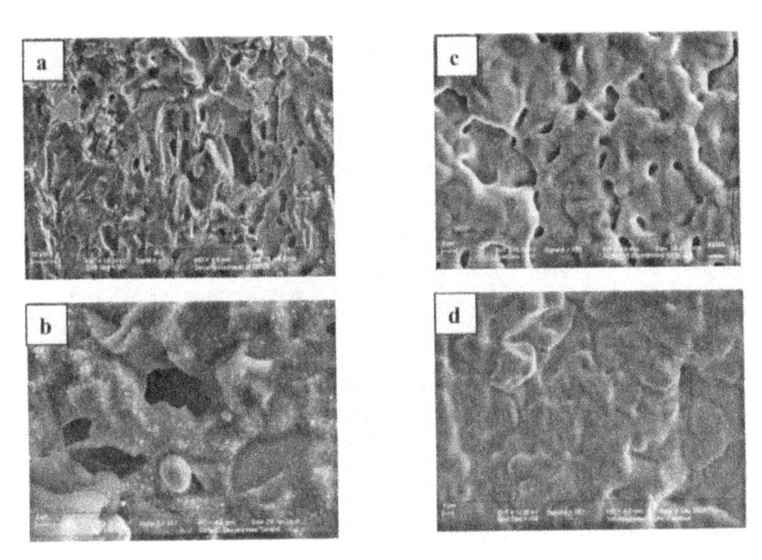

Fig. 2: Scanning electron micrographs of the surface microstructures of vacuum-decomposed MAX phases; (a) Ti_2AlN, (b) Ti_4AlN_3, (c) Ti_3SiC_2 and (d) Ti_3AlC_2.

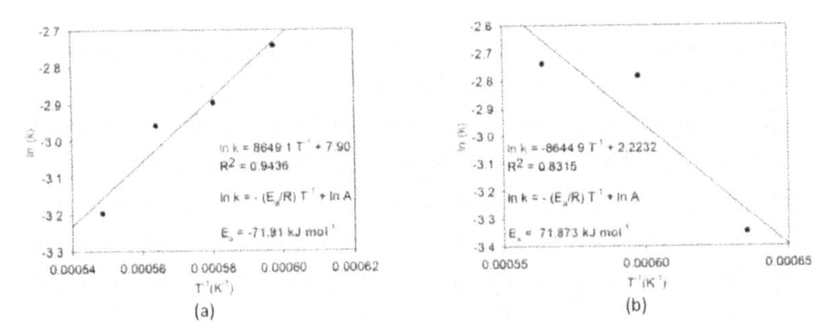

Fig. 3: Calculation of activation energy from the Arrhenius equation for (a) bulk Ti_3AlC_2 and (b) powdered Ti_3AlC_2.

CONCLUSIONS

MAX phases are susceptible to thermal dissociation at 1300-1550 °C in high vacuum. Above 1400 °C, MAX phases decomposed to binary carbide (e.g. TiC_x) or binary nitride (e.g. TiN_x), primarily through the sublimation of A-elements such as Al or Si, which results in a porous surface layer of MX_x being formed. Except for bulk Ti_3AlC_2 which exhibited negative activation energy, positive activation energies were determined for the decomposition of other bulk MAX phases. The pore microstructure of decomposed MAX phase has been shown to play a vital role in the kinetics of decomposition with coarse-pores facilitating the decomposition process but the fine-pores hindering it.

ACKNOWLEDGEMENTS

This work was funded by an ARC Discovery-Project grant (DP0664586) and an ARC Linkage-International grant (LX0774743). Neutron beamtime at ISIS (RB920121) and OPAL was provided by the Science and Technology Facilities Council together with financial support from AMRFP, AINSE (08/329 & 09/606) and LIEF grants (LE0882725). We are grateful to our colleague E/Prof. B. O'Connor for guidance on Rietveld analysis and wish to thank Dr. R. Smith of ISIS, and both Dr. V. Peterson & Dr. S. Kennedy of the Bragg Institute at ANSTO for technical assistance in data collection.

REFERENCES:

1. M.W. Barsoum, The $M_{n+1}AX_n$ phases: A New Class of Solids: Thermodynamically Stable Nanolaminates, *Prog. Solid State Chem.* **28**, 201 (2000).
2. M.W. Barsoum and T. El-Raghy, The MAX Phases: Unique New Carbide and Nitride Materials, *Am. Sci.* **89**, 334-343 (2001).
3. W. Tian, *et al.*, Synthesis and Characterization of Cr_2AlC Ceramics Prepared by Spark Plasma Sintering, *Mater. Lett.* **61**, 4442 (2007).
4. M.W. Barsoum and T. El-Raghy, Synthesis and Characterisation of a Remarkable Ceramic: Ti_3SiC_2, *J. Am. Ceram. Soc.* **79**, 1953 (1996).
5. I.M. Low, *et al.*, Contact Hertzian Response of Ti_3SiC_2 Ceramics, *J. Am. Ceram. Soc.* **81**, 225 (1998).
6. I.M. Low, Vickers Contact Damage of Micro-Layered Ti_3SiC_2, *J. Eur. Ceram. Soc.* **18**, 709 (1998).
7. I.M. Low and Z. Oo, Diffraction Studies of a Novel Ti_3SiC_2–TiC System with Graded Interfaces, *J. Aust. Ceram. Soc.* **38**, 112-115 (2002).
8. I.M. Low, Depth-Profiling of Phase Composition in a Novel Ti_3SiC_2–TiC System with Graded Interfaces, *Mater. Lett.* **58**, 927-930 (2004).
9. I.M. Low and W.K. Pang, Kinetics of Decomposition in MAX Phases at Elevated Temperature, *Mater. Aust. Mag.* **6**, 33-35 (2011).
10. I.M. Low, Z. Oo and K.E. Prince, Effect of Vacuum Annealing on the Phase Stability of Ti_3SiC_2, *J. Am. Ceram. Soc.* **90**, 2610 (2007).
11. Z. Oo, I.M. Low and B.H. O'Connor, Dynamic Study of the Thermal Stability of Impure Ti_3SiC_2 in Argon and Air by Neutron Diffraction, *Physica B*, **385-386**, 499-501 (2006).
12. W.K. Pang, I.M. Low, B.H. O'Connor, A.J. Studer, V.K. Peterson and J.P. Palmquist, Diffraction Study of High-Temperature Thermal Dissociation of Maxthal Ti_2AlC in Vacuum, *J. Alloys Compds.* **509**, 172-176 (2010).
13. W.K. Pang, I.M. Low, B.H. O'Connor, A.J. Studer, V.K. Peterson, Z.M. Sun and J-P Palmquist, Comparison of Thermal Stability in MAX 211 and 312 Phases, *J. Physics: Conference Series*, **251**, 012025 (2010).
14. W.K. Pang and I.M. Low, Diffraction Study of Thermal Dissociation in the Ternary Ti-Al-C System, *J. Aust. Ceram. Soc.* **45**, 39-43 (2009).
15. W.K. Pang, I.M. Low and Z.M. Sun, In-Situ High-Temperature Diffraction Study of Thermal Dissociation of Ti_3AlC_2 in Vacuum, *J Am. Ceram. Soc.* **93**, 2871-2876 (2010).
16. I.M. Low, W.K. Pang, S.J. Kennedy and R.I. Smith, High-Temperature Thermal Stability of Ti_2AlN and Ti_4AlN_3: A Comparative Diffraction Study, *J. Eur. Ceram. Soc.* **31**, 159-166 (2011).
17. H. Zhang, *et al.*, Titanium Silicon Carbide Pest Induced by Nitridation, *J. Am. Ceram. Soc.* **91**, 494 (2008).
18. A. Feng, T. Orling and Z.A. Munir, Field-Activated Pressure-Assisted Combustion Synthesis of Polycrystalline Ti_3SiC_2, *J. Mater. Res.* **14**, 925 (1999).
19. N.F. Gao, Y. Miyamoto and D. Zhang, On Physical and Thermochemical Properties of High-Purity Ti_3SiC_2, *Mater. Lett.* **55**, 61 (2002).
20. J. Emmerlich, *et al.*, Thermal Stability of Ti_3SiC_2 Thin Films, *Acta Mater.* **55**, 1479 (2007).
21. M.W. Barsoum and T. El-Raghy, Synthesis and Characterization of Remarkable Ceramic: Ti_3SiC_2, *J. Am. Ceram. Soc.* **79**, 1953 (1996).

22. N. Tzenov, M.W. Barsoum and T. El-Raghy, Influence of Small Amounts of Fe and V on the Synthesis and Stability of Ti_3SiC_2, *J. Eur. Ceram. Soc.* **20**, 801 (2000).
23. R. Radakrishnan, *et al.*, Synthesis and High-Temperature Stability of Ti_3SiC_2, *J. Alloys Compd.* **285**, 85 (1999).
24. A.K. Galwey and M.E. Brown, Application of Arrhenius Equation to Solid State Kinetics: Can this be Justified?, *Thermochimica Acta.* **386**, 91-98 (2002).
25. H.B. Callen, *Thermodynamics and An Introduction to Thermostatistics.* Wiley, 1985.
26. www.veeco.com/library/Learning_Center/Growth_Information/Vapor_Pressure_Data_For_Selected_Elements/index.aspx.

THE HARDNESS OF ZrB_2 BETWEEN 1373 K AND 2273 K

Jianye Wang, E. Feilden-Irving, Luc J. Vandeperre
Department of Materials & Centre for Advanced Structural Ceramics, Imperial College London, South Kensington Campus, London SW7 2AZ, UK
F. Giuliani
Department of Mechanical Engineering / Department of Materials & Centre for Advanced Structural Ceramics, Imperial College London, South Kensington Campus, London SW7 2AZ, UK

ABSTRACT

There are not that many materials that can be considered for applications where very high temperatures (>2273 K) are expected. Zirconium diboride (ZrB_2) is one of a group of materials, whose melting point makes them potential candidates. Despite the intention of using this material at elevated temperatures, most reports on mechanical properties have focussed on room temperature, with only very few results obtained at very high temperatures. Since indentation testing can give an indication of yield strength and creep resistance, the aim of this work was to develop a technique for determining the hardness of ZrB_2 at very high temperatures (up to 2273 K). The technique consists of pushing two wedges of the material onto each other so that the contact area increases until equilibrium is reached. To validate the approach, measurements were first made on aluminium oxide specimens. Its variation of hardness with temperature was found to be in reasonable agreement with literature data. The experiments on ZrB_2 show that its hardness decreases from 6.2 GPa at 1373 K to 0.61 GPa at 2273 K.

INTRODUCTION

The number of compounds with very high melting points is limited and therefore not that many materials can be considered for applications where very high temperatures will be encountered. Examples of such applications are leading edges of hypersonic airplanes or next generation space re-entry vehicles with increased manoeuvrability. Zirconium diboride (ZrB_2) is one of a group of materials, whose melting point makes them potential candidates. Other advantages of ZrB_2 are that it is relatively inexpensive and easier to process than many other candidate materials[1-10]. A weak point of ZrB_2 is that it is susceptible to active oxidation[11-14].

Despite the intention of using this material at very high temperatures (>2273 K), most reports on mechanical properties have been room temperature measurements, with only very few results obtained at moderate to high temperatures: the variation of the elastic properties has been measured up to 1373 K[15], a range of strength and sometimes toughness measurements have been carried at temperatures up to 2073 K[6,16], and the hardness of ZrB_2 or its composites has been measured up to 1273 K[17-20]. Hence, there is a need to improve the understanding of the mechanical behaviour of these materials at the intended service temperatures.

In addition to the value of the hardness, analysis of indentation data can also yield information on the flow stress and yield behaviour[21-23] or resistance to creep[24-29]. As indentation experiments are normally relatively easy to carry out, they appear attractive for high temperature testing. However, very high temperature indentation is difficult. One problem is that diamond tips can't be used because diamond is susceptible to either oxidation or graphitisation. Moreover, as the temperature increases, reactions between diamond and the material being indented become increasingly more difficult to avoid. Replacing diamond with other materials for testing hard materials is difficult as the hardness of alternatives such as sapphire decreases rapidly with temperature[30]. Another alternative sometimes considered is cubic boron nitride, but like diamond this material suffers from oxidation and can react with the material being tested. To circumvent the problem of reactions between the indenter and the material being tested entirely, Atkins and Tabor[31] developed a cross-bar technique in which two samples of the same material are pressed together. In such tests, the hardness is not defined as the ratio of force and area of residual imprint but as the ratio of force and the area of mutual contact that forms during the test. To ensure that this equilibrium area can be found in a single test, they used either cylindrical or wedge shaped samples so that the contact area can increase as long as the stress is high enough for the samples to deform each other. The aim of this paper was therefore to use this approach to measure the hardness of zirconium diboride at very high temperatures (between 1373 K and 2273 K). To understand how the hardness obtained in such tests compares with other techniques, the same technique was also used to measure the variation of the hardness of aluminium oxide in the range 1273 K to 1773 K. Aluminium oxide was selected because hardness data for aluminium oxide obtained using high temperature indentation with diamond tips is available in the literature[30].

EXPERIMENTAL

It was decided to produce pure ZrB_2 samples because earlier work had indicated that the presence of other phases can potentially influence the hardness values obtained[19]. To ensure sufficient densification, the ZrB_2 was produced by hot pressing a commercial ZrB_2 powder (Grade B , H.C. Stark, Germany) without additives at 2273 K for 1.3 h under a pressure of 30 MPa. Al_2O_3 samples were obtained by slip casting a commercial Al_2O_3 powder (CT3000 SG, Almatis, Germany). The slip contained 30 wt% Al_2O_3 in demineralised water and was stabilised by adding a commercial dispersant (Dolapix CA , Zschimmer & Schwarz, Germany). After casting and drying, the samples were sintered in air at 1873 K for 1 h. The density of the materials was determined using Archimedes' principle.

For hardness measurements at elevated temperature, use was made of a high temperature dilatometer (DIL 402C, Netzsch, Germany) in which contact between the push-rod and the sample is maintained by exerting a small compressive force on the sample with the aid of a spring in the cold part of the dilatometer. The fact that the spring is in the cold part of the dilatometer is important since this means that the force exerted is independent of temperature. To determine the force exerted by the push-rod, a calibration was carried out by allowing a Berkovich diamond indenter to indent a polished aluminium sample and comparing the area of the imprint with a separately obtained calibration of the variation of the area of the imprint with applied load. The latter was obtained using the micro-head of an instrumented nano-indenter (Nanotest Platform 2, Micro Materials Ltd, United Kingdom). Hardness measurements were then made by placing two wedge-shaped samples in a cross-bar configuration in the dilatometer as shown in Figure 1.

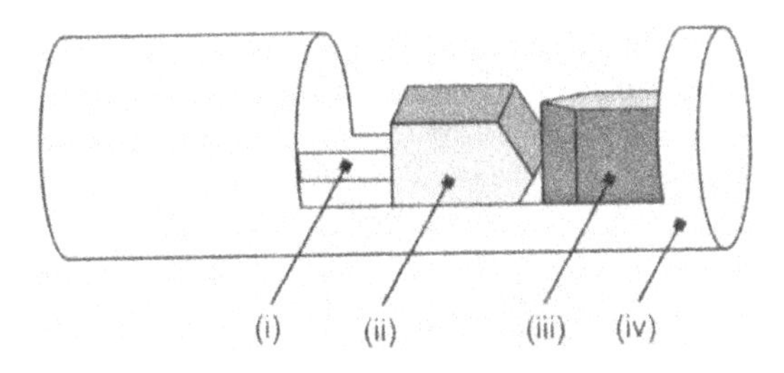

Figure 1. Configuration of the samples in the dilatometer: (i) the push-rod of the dilatometer, (ii) sample A, (iii) sample B, and (iv) the dilatometer tube.

The dilatometer was then heated to the test temperature (10 K h^{-1}), held for 0.5 h, followed by rapid cooling. The included angle of the wedges was 120° for both materials. The measurements were carried out in a graphite tube with a graphite push-rod under flowing He.

RESULTS

Both materials were essentially fully dense (>99%).

Figure 2 shows the calibration line of the area of the indents versus load obtained by making Berkovich indents using an instrumented indenter and measuring the area of the indents on scanning electron micrographs. From this calibration line, the hardness of the aluminium is estimated to be 1.33 ± 0.06 GPa, which agrees with typical values obtained for the small scale hardness of Al in the literature[32]. Also shown in Figure 2 is that the average area and the 99% confidence band for 9 imprints formed by allowing the pushrod to press the same indenter into the same aluminium sample. From this the load exerted by the spring is determined with 99% certainty to be within the range of 1177 mN to 1847 mN, with a best estimate of 1498 mN. The latter value was used to determine the hardness in further experiments.

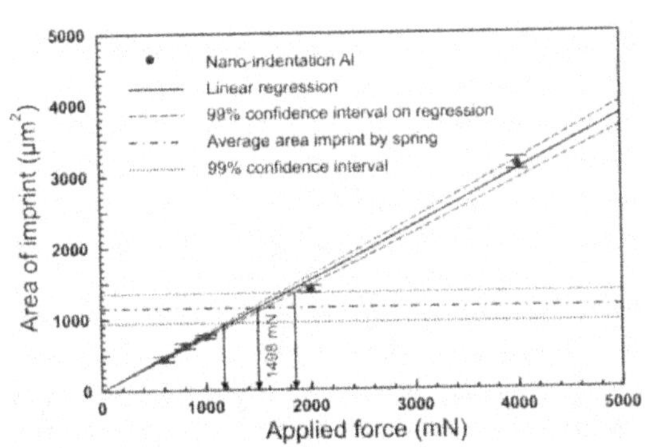

Figure 2. Linear regression and 99% confidence band for the area of indents versus load for a Berkovich diamond indenter pressed against aluminium as well as the average and 99% confidence bands for the area of the indents made by allowing the spring to indent the Al.

Figure 3 shows an example of the deformed contact area after heating to 1773 K under constant load for Al_2O_3. The imprint is not entirely symmetrical about the centre-line indicating that misalignments can form. Figure 4 compares the hardness measurements obtained with the cross-bar technique with values obtained by Vickers indentation with 0.5-1 N load taken from the work of Westbrook[30], Alpert et al.[34] and Koester and Moak[35]. Unfortunately, it proved impossible to generate data at lower temperatures to compare with more of the literature data due to the fact that the size of the imprint became too small for the fixed applied load. But the data obtained by cross bar does appear to agree reasonably with other high temperature data obtained with a boron carbide Vickers indenter in reducing atmosphere.

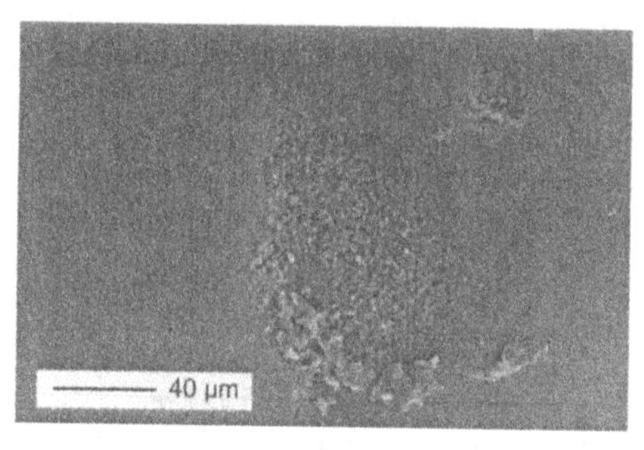

Figure 3. Example of an imprint formed during self-indentation of Al_2O_3 at 1673 K.

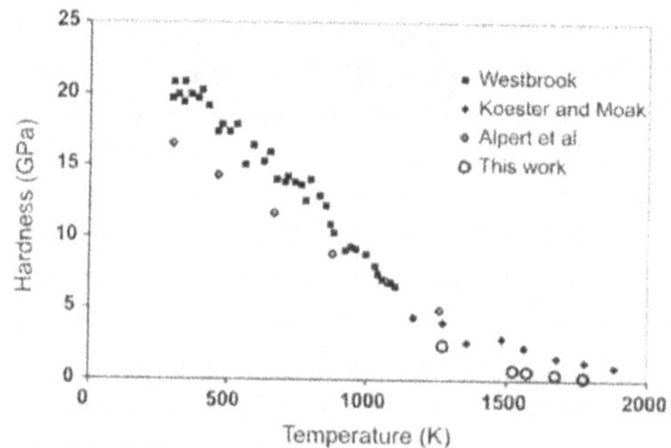

Figure 4. Comparison of the hardness values obtained here using the cross-bar technique with literature data for Al2O3 obtained by Vickers indentation[30,34-35]

Figure 5 shows examples of the imprints formed during experiments on ZrB_2. At all temperatures the area of the mutual imprint can easily be recognised because the intimate contact between the two materials at high temperature leads to diffusion bonding. The bond fractures during cooling or removal of the samples, leaving a clearly distinguishable mark on the surface of the samples. At 1873 K, a regular diamond shape imprint is formed with limited additional features except for some evidence of limited cracking as indicated with the arrow. At 2073 K, the shape of the indent suggests that one sample had slightly tilted relative to the other one. As a result, in the region indicated

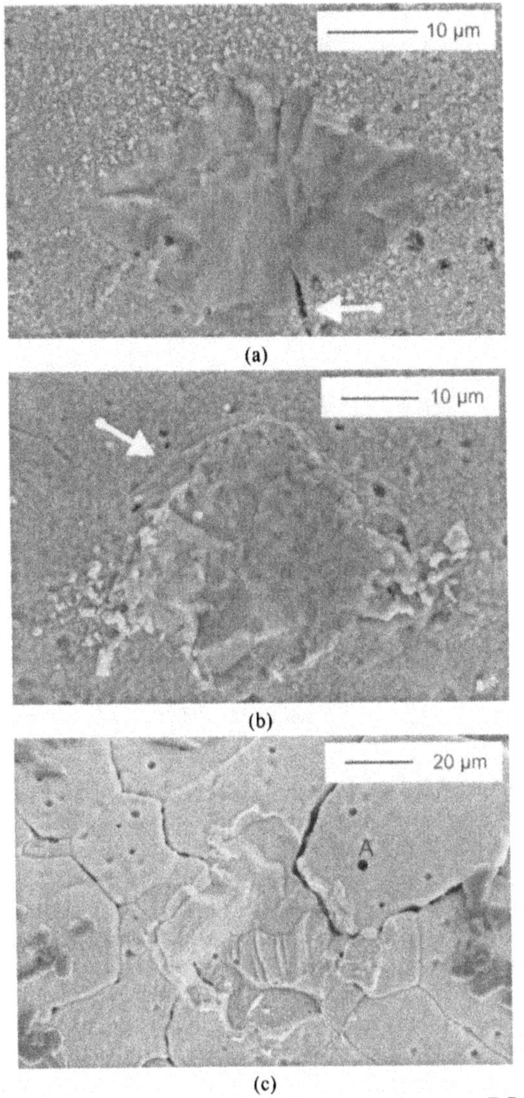

(c)

Figure 5. Examples of the imprints formed during experiments on ZrB_2 at: (a) 1873 K, (b) 2073 K , and (c) 2273 K.

by the arrow, material has been pushed outward creating clear ripples in the surface, which are clear evidence of some ductility. At 2273 K, the material is now so soft that the imprint becomes larger than the area taken up by single grains and as illustrated by the behaviour of the grain marked "A" in the photograph, this leads to detachment of grains by fracturing along the grain boundaries in response to the levering action of the opposite wedge.

The results of the hardness measurements are compared with literature information for lower temperatures in Figure 6. Obviously, the hardness is much lower at very high temperatures than it is at room temperature. However, because the measurements obtained at the lowest temperatures of the high temperature experiments carried out here tend towards the values measured up to 1273 K in the literature, it appears that there are three distinct regions for the variation of the hardness of ZrB_2 with temperature: a rapid decline in hardness near room temperature, a region where there is limited variation in hardness with temperature and a third region where the hardness decreases rapidly again.

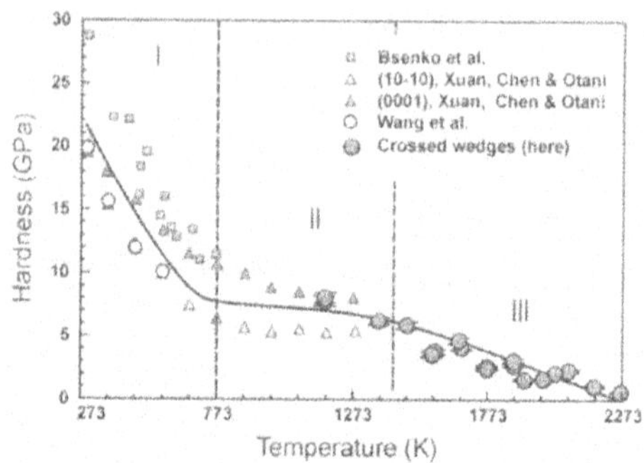

Figure 6. Results of hardness measurements of ZrB_2 from 1173 K to 2273 K compared with published information on the temperature dependence of its hardness taken from[17-20]. The solid line is merely a suggested trend.

DISCUSSION

The scanning electron micrographs of the mutual indentation shapes illustrates that hardness imprints caused by the mutual indentation of the same material adopt a form largely consistent with what can be expected for plastic deformation when two wedges are pressed against each other in a crossed alignment. Although there is some evidence of fracture playing a role, either when the measurement temperature is low (white arrow in Figure 5a) or when the indent size is similar to the grain size (see grain A in Figure 5c), plastic deformation, i.e. material flow, plays a key role in the formation of the imprints. Moreover, the bonding that occurs between the two samples in close contact aids in recognizing the correct indentation size as the morphology of the fracture surface is very different from that of the surrounding material. Moreover, the fact that the data for Al_2O_3 obtained here aligns reasonably with the available literature information, further supports the idea that hardness measurements with crossed wedges form an elegant tool to widen the data basis at very high temperatures and for materials, which are likely to react with a range of possible indenter materials.

The high temperature data for ZrB_2 obtained in this work complements work to more limited temperatures and can aid in providing an interpretation of the variation of the hardness with temperature. Near room temperatures, the hardness of ZrB_2 drops rapidly, consistent with an interpretation that it is the lattice resistance to dislocation motion that controls the hardness in this region as discussed in detail by Wang et al.[19]. Above ~773 K, however, single crystal hardness measurements suggested that the hardness would become constant. The measurements made here at much higher temperatures, seem to support this observation: the hardness at much higher temperatures is more consistent with an intermediate temperature region of limited hardness change than with a continued decrease in hardness as near room temperature. Hence, this means that from above 773 K onwards, the hardness is no longer controlled by the lattice resistance but by some other obstacles to dislocation flow. Since the material under investigation was a relatively coarse grained ZrB_2 with negligible second phases within the grains, a prime candidate is the presence of other dislocations (forest dislocations). The latter are obstacles with a much higher activation energy than the lattice resistance[33] so that thermal energy is not sufficient to overcome these obstacles leading to a much more a-thermal hardness. At very elevated temperatures, however, the hardness does decrease again with temperature albeit less dramatically than near room temperature. Possible reasons for the further decrease in hardness are the operation of typical creep mechanisms: dislocation flow controlled by climb over the obstacles or creep caused by diffusion flow of material (Cobble or Nabarro-Herring creep). All these processes typically require diffusion and hence can only occur at sufficient rate at very elevated temperatures. Hence, the second drop in hardness only occurs when such high temperatures (~1473 K) are reached.

CONCLUSIONS

Experiments using a cross-bar technique developed by Atkins and Tabor were carried out to measure the hardness of aluminium oxide and zirconium diboride at elevated temperatures.

For Al_2O_3, the hardness values obtained here are in line with an extrapolation of the literature values suggesting the technique gives reasonable estimates for the hardness.

The variation of the hardness of zirconium diboride versus temperature can be divided into three regions: near room temperature, the hardness of ZrB_2 decreases rapidly due to a decrease in the lattice resistance to dislocation motion, at intermediate temperatures, 773-1673 K, the hardness is more or less constant at ~6 GPa. It is suggested that this constant region is a consequence of dislocation flow being limited by obstacles to dislocation motion such as other dislocations generated during the deformation process. At higher temperatures, the hardness decreases again to a value as low as 0.61 GPa at 2273 K. It is suggested that the latter decrease is due to deformation mechanisms normally referred to as creep: contributions of diffusion to the deformation either by assisting climb of dislocations to escape from obstacles or by diffusive flow of matter to relieve the stress.

ACKNOWLEDGEMENTS

The authors would like to thank the Engineering and Physical Sciences Research Council of the UK for funding for this work.

REFERENCES

1. Monteverde, F., Beneficial effects of an ultra-fine alpha-SiC incorporation on the sinterability and mechanical properties of ZrB2. *Applied Physics a-Materials Science & Processing*, 2006. **82**(2): p. 329-337.
2. Zhu, S.M., W.G. Fahrenholtz, and G.E. Hilmas, Influence of silicon carbide particle size on the microstructure and mechanical properties of zirconium diboride-silicon carbide ceramics. *Journal of the European Ceramic Society*, 2007. **27**(4): p. 2077-2083.
3. Zhang, S.C., G.E. Hilmas, and W.G. Fahrenholtz, Pressureless densification of zirconium diboride with boron carbide additions. *Journal of the American Ceramic Society*, 2006. **89**(5): p. 1544-1550.
4. Medri, V., et al., Comparison of ZrB2-ZrC-SiC composites fabricated by spark plasma sintering and hot-pressing. *Advanced Engineering Materials*, 2005. **7**(3): p. 159-163.
5. Chamberlain, A.L., et al., High-strength zirconium diboride-based ceramics. *Journal of the American Ceramic Society*, 2004. **87**(6): p. 1170-1172.
6. Guo, S.Q., Densification of ZrB_2-based composites and their mechanical and physical properties: a review. *Journal of The European Ceramic Society*, 2009. **29**: p. 995-1011.
7. Zhang, S.C., G.E. Hilmas, and W.G. Fahrenholtz, Pressureless sintering of ZrB_2-SiC ceramics. *Journal of the American Ceramic Society*, 2008. **91**(1): p. 26-32.
8. Zhu, S.M., Pressureless sintering of carbon-coated zirconium diboride powders. *Materials science & engineering. A, Structural materials*, 2007. **459**(1-2): p. 167-171.
9. Kuzenkova, M.A. and P.S. Kislyi, Grain growth in zirconium diboride during sintering. *Powder metallurgy and metal ceramics*, 1966. **5**(1): p. 10-13.
10. Kislyi, P.S. and M.A. Kuzenkova, Regularities of sintering of zirconium diboride-molybdenum alloys. *Poroshkovaya Metallurgiya*, 1966. **5**: p. 16-23.
11. Karlsdottir, S.N. and J.W. Halloran, Rapid oxidation characterization of ultra-high temperature ceramics. *Journal of the American Ceramic Society*, 2007. **90**(10): p. 3233-3238.
12. Sciti, D., M. Brach, and A. Bellosi, Long-term oxidation behavior and mechanical strength degradation of a pressurelessly sintered ZrB_2-$MoSi_2$ ceramic. *Scripta Materialia*, 2005. **53**: p. 1297-1302.
13. Karlsdottir, S. and J.W. Halloran, Zirconia transport by liquid convection during oxidation of zirconium diboride-silicon carbide. *Journal of the American Ceramic Society*, 2008. **91**(1): p. 272-277.
14. Opila, E., S. Levine, and J. Lorincz, Oxidation of ZrB_2- and HfB_2-based ultra-high temperature ceramics: effect of Ta additions. *Journal of Materials Science*, 2004. **39**(19): p. 5969-5977.
15. Okamoto, N.L., et al., Temperature dependence of thermal expansion and elastic constants of single crystals of ZrB_2 and the suitability of ZrB_2 as a substrate for GaN film. *Journal of Applied Physics*, 2003. **93**(1): p. 88-93.
16. Hu, P. and Z. Wang, Flexural strength and fracture behaviour of ZrB_2-SiC ultra-high temperature ceramic composites at 1800 °C. *Journal of The European Ceramic Society*, 2010. **30**(4): p. 1021-1026.
17. Xuan, Y., C.-H. Chen, and S. Otani, High temperature microhardness of ZrB_2 single crystals. *J. Phys. D. : Appl. Phys.*, 2002. **35**: p. L98-L100.
18. Wang, J., F. Giuliani, and L.J. Vandeperre, The effect of load and temperature on hardness of ZrB2 composites. *Ceram. Eng. Sci. Proc.*, 2010. **31**(2): p. 59-68.
19. Wang, J., F. Giuliani, and L. Vandeperre, Temperature and strain-rate dependent plasticity of ZrB2 composites from hardness measurements. *Ceram. Eng. Sci. Proc.*, 2011. **in press**.

20. Bsenko, L. and T. Lundström, The high-temperature hardness of ZrB_2 and HfB_2. *Journal of the Less Common Metals*, 1974. **34**(2): p. 273-278.
21. Vandeperre, L.J., F. Giuliani, and W.J. Clegg, Effect of elastic surface deformation on the relation between hardness and yield strength. *Journal of Materials Research*, 2004. **19**(12): p. 3704-3714.
22. Cheng, Y.T. and C.M. Cheng, Can Stress-Strain Relationships be Obtained from Indentation Curves using Conical and Pyramidal Indenters. *J. Mat. Res.*, 1999. **14**(9): p. 3493-3496.
23. Lee, B.W., et al. Determining Stress-strain Curves for Thin Films by Experimental/Computational Nanoindentation. in *MRS Fall Meeting*. 2004. Boston.
24. Asif, S.A.S. and J.B. Pethica, Nanoindentation creep of single-crystal tungsten and gallium arsenide. *Philosophical Magazine a-Physics of Condensed Matter Structure Defects and Mechanical Properties*, 1997. **76**(6): p. 1105-1118.
25. Poisl, W.H., W.C. Oliver, and B.D. Fabes, The Relationship between Indentation and Uniaxial Creep in Amorphous Selenium. *J. Mater. Res.*, 1995. **10**(8): p. 2024-2032.
26. Yue, Z.F., J.S. Wan, and Z.Z. Lu, Determination of creep parameters from indentation creep experiments. *Applied Mathematics and Mechanics-English Edition*, 2003. **24**(3): p. 307-317.
27. Fischer-Cripps, A.C., A simple phenomenological approach to nanoindentation creep. *Materials Science and Engineering*, 2004. **A385**: p. 74-82.
28. Brookes, C.C.A., R.P. Burnand, and J.E. Morgan, Anisotropy and indentation creep in crystals with the rocksalt structure. *Journal of materials science*, 2002. **10**(Volume 10, Number 12 / December, 1975): p. 2171-2173.
29. Goodall, R. and T.W. Clyne, A critical appraisal of the extraction of creep parameters from nanoindentation data obtained at room temperature. *Acta materialia*, 2006. **54**(20): p. 5489-5499.
30. Westbrook, J.H., The temperature dependence of hardness of some common oxides. *Rev. Hautes Temper. et Refract.*, 1966. **3**(1): p. 47-57.
31. Atkins, A.G., C.A. Brookes, and M.J. Murray, Recent Studies of the Mechanical Properties of Solids at High Temperatures. *Rev. Hautes Temper. et Refract.*, 1966. **3**(1): p. 19-25.
32. Tsui, T.Y., W.C. Oliver, and G.M. Pharr, Influences of stress on the measured mechanical properties using nanoindentation : Part I. Experimental studies in an aluminium alloy. *Journal of Materials Research*, 1996. **11**(5): p. 752-759.
33. Frost, H.J. and M.F. Ashby, Deformation mechanism maps: the plasticity and creep of metals and ceramics. 1982, Oxford: Pergamon Press.
34. Alpert, C.P., et al., Temperature Dependence of Hardness of Alumina-based Ceramics. *Journal of the American Ceramic Society*, 1988. 71(8): p. C371-C373.
35. Koester, R.D. and D.P. Moak, Hot hardness of selected borides, oxides and carbides to 1900°C. *Journal of the American Ceramic Society*, 1967. 50(6): p. 290-296.

MICROSTRUCTURAL ANALYSIS OF A C_f/ZrC COMPOSITES PRODUCED BY MELT INFILTRATION

Linhua Zou[a], Sergey Prikhodko[a], Timothy Stewart[b], Brian Williams[b] and Jenn-Ming Yang[a]

[a] Department of Materials Science and Engineering, University of California Los Angeles
Los Angeles, CA 90095, USA

[b] Ultramet, Pacoima
Pacoima, CA 91331, USA

ABSTRACT

The microstructure of a reactive melt infiltrated C/ZrC composite by using a Zr-10at%Si alloy was investigated by XRD and TEM. The phase in the composite was first analyzed by XRD, and phases in the matrix were further confirmed by SEM/EDS, STEM/EDS and TEM. It demonstrates that there exist α-Zr, Zr_3Si and ZrC in the matrix of the composites.

INTRODUCTION

Ultrahigh temperature materials (UHTMs) have been identified as one of the critical enabling technologies for hypersonic flight, atmospheric re-entry and rocket. Among all the potential refractory materials, continuous carbon fiber reinforced ZrC composite is one of the most promising materials for ultrahigh temperature structural applications. ZrC possesses a melting point as high as 3540 °C, the oxidation resistance of ZrC is comparable to hafnium carbide. However, the density of ZrC (6.73 g/cm^3) is much lower than that of the HfC (12.2 g/cm^3), that makes ZrC more appealing for lightweight rocket components. Like most ceramics, ZrC is brittle and therefore its applications remain limited in its monolithic form. Long continuous carbon fibers are incorporated into ZrC to improve its fracture resistance. Such fiber-reinforced ZrC matrix composites have been explored as materials possessing structural characteristics suitable for ultra high temperature applications in thermal protection systems. The manufacture of C_f/ZrC by some widely used methods, such as hot pressing, sparking plasma sintering and chemical vapor infiltration, present shape, size and density limitations. Reactive melt infiltration (RMI) processing has been successfully applied to the Si-C system to fabricate C_f/SiC and SiC/SiC composites [1-5]. This same processing technique has also been successfully used in recent years to fabricate C_f/ZrC composites [6-7]. Just as Si-Mo alloy melt was used for reactive infiltration in Si-C system to reduce residual silicon and incorporate silicate compound into SiC ceramics [3-5], Zr-Si alloy melt was used in Zr-C system for similar purpose. In this paper, microstructural analysis of the C_f/ZrC composite fabricated by RMI of a Zr-10 at% Si alloy was conducted, and the phase in the composites was studied by XRD, SEM/EDS, STEM/EDS, and TEM.

EXPERIMENTAL

Materials Preparation

The carbon fiber-reinforced ZrC composite was fabricated by Ultramet (Pacoima, CA) using a proprietary RMI processing method. Carbon fiber preform was laid up by hand from 2D plain weave T300 fabric. The fiber preform was first coated with a pyrolytic carbon interface layer by chemical vapor infiltration (CVI). Next, a controlled level of carbon was deposited onto the preform by CVI to form a porous C/C skeleton. Molten Zr-10 at % Si alloy then infiltrated the porous preform by wicking action and was drawn along the carbon fiber tows by capillary forces, where it reacted with the deposited carbon to form ZrC.

Materials Characterization

The cross section of the C_f/ZrC composite sample was polished to 0.05μm for FIB TEM sample preparation, phase analysis by XRD, SEM, EDS and TEM characterization. The TEM sample preparation was conducted by using FEI Nova 600 DualBeam™- SEM/FIB system (5350 NE Dawson Creek Drive, Hillsboro, Oregon 97124 USA). Its SEM resolution is 1.1nm, and FIB resolution 10nm.

An X-ray powder diffractometer was used to carry out phase analysis using a Cu target. An X'celerator RTMS Scanning Detector (PANalytical, Almelo, The Netherlands) was used, with voltage was set as 45 kV and current as 40 mA. X'pert highscore software was used to identify the peaks.

Nova™ NanoSEM 230 was also used to perform image analysis. The system is the most versatile ultra-high resolution FESEM combining very low kV imaging and analytical capabilities. With SE/BSE in-lens detection and beam deceleration technologies, the Nova NanoSEM 230 is a complete solution for ultra-high resolution characterization at high and low voltage in high vacuum.

TEM studies were conducted by using Titan 300kv S/TEM (FEI). With a field emission gun (FEG), this microscope is designed for both high resolution TEM/STEM and analytical microscopy. It has three attachments: (1) A scanning unit with a high-angle annular dark-field (HAADF) detector for scanning TEM (Z-contrast imaging); (2) Electron dispersive X-ray analysis system for chemical information; (3) Gatan wide-angle CCD (2kx2k) for image recording. Point resolution is 0.2nm, and HR- STEM resolution 0.136nm. Double tilt holder was used for loading sample.

RESULTS AND DISCUSSION

XRD Phase Analysis

As shown in Figure1, XRD pattern from the cross sectional surface of the sample shows that there exist four phases in the material, i.e. zirconium, carbon, zirconium carbide, and Zr_3Si. The carbon is mainly from carbon fiber bundle and un-reacted pyrolytic carbon deposited by CVD. The result indicates that residual metal zirconium still remain in the final composite, the reaction between molten zirconium and pyrolytic carbon happened during RMI processing and produced zirconium carbide, and Zr_3Si precipitated from molten Zr-10at%Si alloy as temperature was cooling down. The formation of Zr_3Si is in agreement with the Zr-Si binary system [8]. Zr_3Si first precipitated as molten Zr-10at% Si alloy underwent cooling, and then the residual liquid alloy transformed into Zr_3Si and β-zirconium at 1570°C through a eutectic reaction; as the temperature continued to drop, the phase transformation of β-Zr→α-Zr occurred at 863°C [8]. However, the XRD results need to be further confirmed.

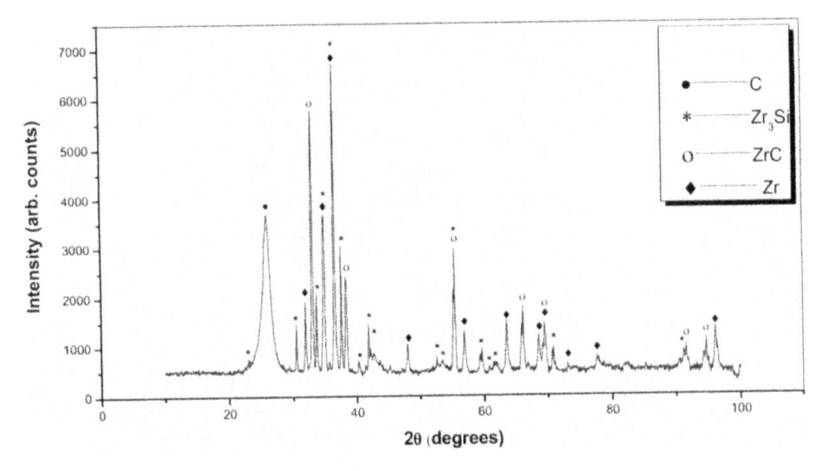

Figure1 X-ray diffraction pattern of the C_f/ZrC composite manufactured by reactive melt infiltration with a 10at%Si-Zr alloy

SEM/EDS Analysis

EDAX analysis was performed on an area which is close to the site where the TEM sample was prepared. In Figure 2a, the typical microstructural features of a C_f/ZrC composite are shown, including continuous phase, precipitated particles, and the continuous layer around carbon fiber bundle. The area in the dashed line frame shown in Figure 2b was chosen as a typical unit for the analysis, and the analyzed points were marked by different numbers (Figure 2c). It is obvious that the whole area contains Zr, Si, and C elements. The results obtained from these points are shown in Table 1. It can be seen that the spectra obtained from the points 1-3 on the big precipitated particle only possess Si and Zr peaks, the molar ratio of Zr to Si is about 3, and the particle is supposed to be Zr_3Si; for the spectra from the points 4-6 on the continuous phase, there is only Zr peak in the spectra from the points 4-5, the corresponding phase is supposed to be zirconium, however in the spectrum from the point 6, it contains carbon and zirconium peaks, the phase might be ZrC according to its molar ratio, C/Zr=0.6. This is possible when a ZrC particle is just beneath the surface of the continuous phase, and contributes to XRD. The spectra from the points 7-9 on the particles close to the carbon fiber bundle indicate there exist zirconium and carbon peaks, the ratio of C to Zr is 1.2~1.6, and so they should be ZrC. The existence of Zr_3Si and Zr corresponds with theoretical analysis based on Zr-Si binary system as 10at % Si is incorporated into Zr [8].

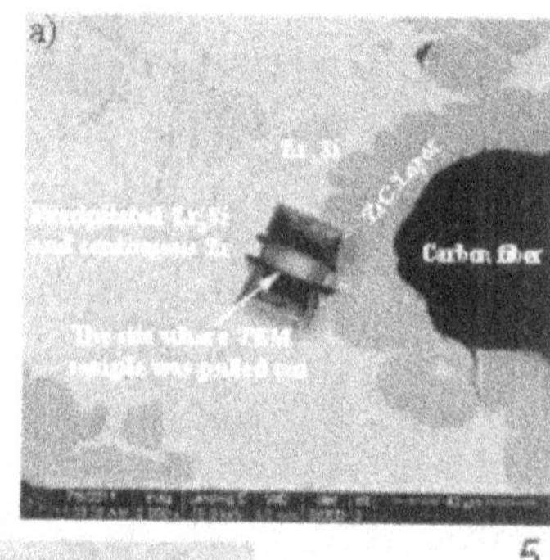

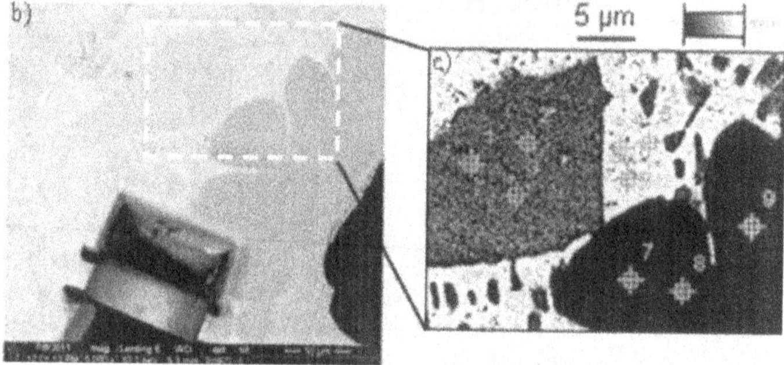

Figure 2 SEM/EDS elemental analysis on different phases, which are close to the site where TEM sample was prepared: a) SEM micrograph showing the typical features in a C_f/ZrC composite fabricated by RMI; b) SEM image showing the site and the rectangular area with dashed line where EDS was conducted; c) magnified SEM image showing the area with analyzed points marked

Table 1 The element composition from EDS analysis (at %)

Points	C-K	Si-K	Zr-L	Molar ratio (Zr/Si or C/Zr)	Possible phases
1	-	25.9	72.1	2.8	Zr_3Si
2	-	25.5	74.5	2.9	Zr_3Si
3	-	25.8	74.2	2.9	Zr_3Si
4	-	-	100.0	-	Zr
5	-	-	100.0	-	Zr
6	37.4	-	62.6	0.6	ZrC
7	54.9	0.0	45.1	1.2	ZrC
8	62.0	0.1	37.9	1.6	ZrC
9	54.5	-	45.5	1.2	ZrC

STEM/EDS Analysis

STEM/EDS analysis was further conducted to confirm phases in the TEM sample. Each phase analyzed by EDS was designated as O_1 to O_7 in Figure3a. The element composition obtained from the analysis is shown in table 2. Three types of spectra were obtained. Besides the peaks of carbon, silicon and zirconium, there are peaks of oxygen and copper. Obviously, the copper peak is from the copper grid where the TEM sample was loaded, the oxygen peak might come from contamination during the infiltration process or from the raw Zr-10at%Si alloy used for melt infiltration. The spectra obtained from O_1, O_2, O_4 and O_6 are similar with the typical one shown in Figure3b, in which the atomic ratio of Zr to Si is around 2.6, and it might be Zr_3Si. The spectra from O3 and O5 are almost same with one of them shown in Figure3c. The main peak is Zr, and the small peak of oxygen is due to contamination. The corresponding phase should be the residual zirconium. The spectrum from O7 is mainly composed of two peaks, as shown in Figure 3d, one is carbon, and the other is zirconium. The atomic ratio of carbon to zirconium is 0.4, so the phase might be ZrC.

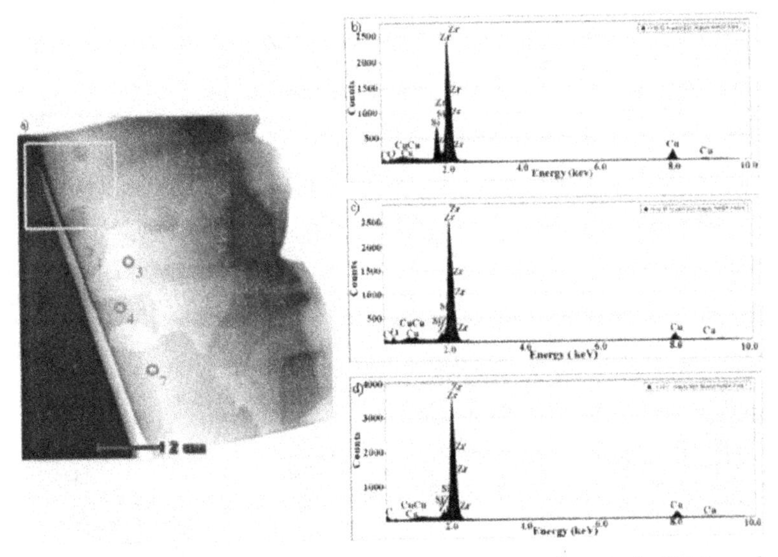

Figure3 S/TEM/EDAX elemental analysis on different grains or phases in the TEM sample (a) STEM image showing the positions (O_1~O_7) where EDAX were performed; (b) the typical spectrum representing those from O_1, O_2, O_4 and O_6 respectively; (c) the typical spectrum representing those from O_3 and O_5 respectively; (d) the typical spectrum from O_7

Table 2 Elemental composition from STEM/EDS analysis (at %)

Points	C	O	Si	Zr	Morlar Ratio Zr/Si or C/Zr	Possible phase
1	1.9	-	27.5	70.6	2.6	Zr_3Si
2	-	-	27.0	73.0	2.7	Zr_3Si
3	-	6.7	-	93.3	-	Zr
4	-	3.7	25.8	70.5	2.7	Zr_3Si
5	-	9.9	-	90.1	-	Zr
6	-	5.2	26.2	68.6	2.6	Zr_3Si
7	26.1	1.6	-	72.2	0.4	ZrC

Electron Diffraction

Selected Area Electron Diffraction (SAED) was done by TEM to further confirm phases mentioned above. The grain where EDS was done at O1 was analyzed by using SAED, the diffraction pattern obtained is shown in Figure 4.The reported crystal structure data of Zr_3Si are a=11.01Å, c=5.45Å, which has the similar structure of Ti_3P [9]. So far, as there is no any XRD data of Zr_3Si reported, it is a challenge to get its diffraction pattern indexed. Based on literature [9] and Ti_3P XRD data, the XRD data of Zr_3Si were obtained through theoretical calculation. Finally the diffraction pattern is indexed to be Zr_3Si along the [011] zone axis. It is in agreement with the results from STEM/EDS analysis (Figure 3b and Table 2), and the analysis from the Zr-Si binary phase diagram [8].

The big grain shown in Figure 5a, where STEM /EDS were taken at O7, was studied by SAED at the circled area, and the obtained diffraction pattern with superlattice spots was indexed to be ZrC along the [011] zone axis (Figure 5b). This further confirms the result from STEM/EDS (Table 2)

The continuous phase among grains was also studied by SAED. The circled area where diffraction was conducted is shown in Figure 6a. The observed reflections in the diffraction pattern are consistent with the crystal structure data of zirconium as viewed along the [2-1-1 0] zone axis (Figure 6b). This further confirms the results from STEM/EDS analysis that the continuous phase is α-Zr (Table 2).

By using TEM, it confirms that there exist Zr_3Si, α-Zr and ZrC phases in the matrix of the C_f/ZrC composites fabricated by RMI. This demonstrates that the composition of the matrix falls into the α-Zr-Zr_3Si-ZrC_{1-x} area in Zr-Si-C ternary system as the matrix is produced through a molten Zr-10at% Si alloy infiltration into a porous carbon fiber preform with certain level of deposited pyrolytic carbon. Meanwhile, in this area, the accompanied silicide compound should be Zr_3Si rather than Zr_2Si, ZrSi, Zr_3Si_2 and Zr_5Si_3. It is worth to note that the Zr_3Si peaks were able to be identified after the phase had been confirmed by TEM.

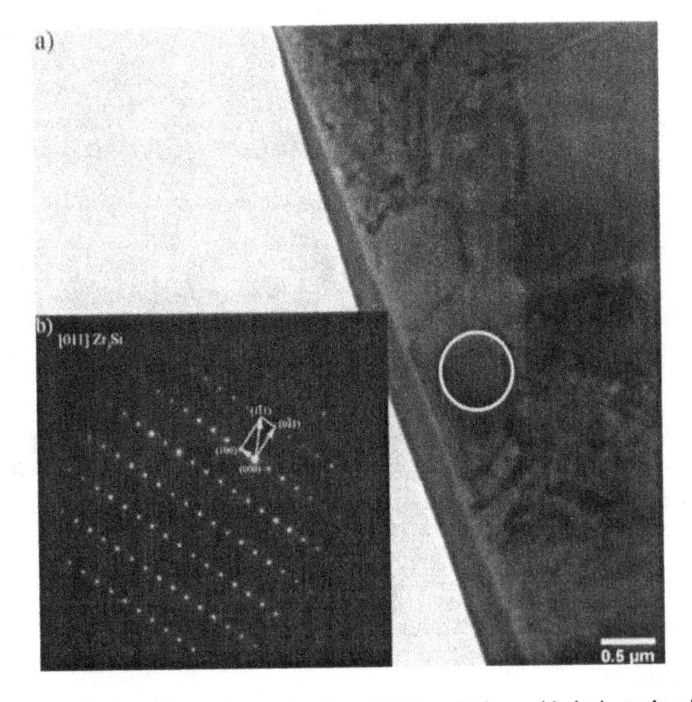

Figure 4 TEM BF image showing the grain where SAED was taken, with the inset showing its corresponding diffraction pattern

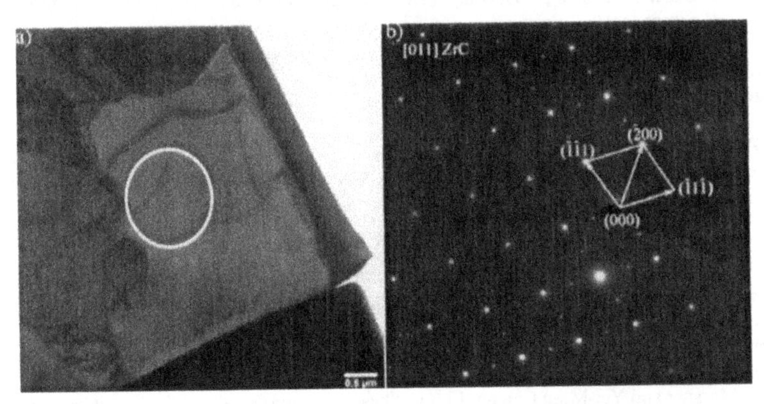

Figure 5 The grain close to the fiber bundle and its SAED pattern: a) TEM BF image showing the grain, and b) the diffraction pattern from the circled area in the grain

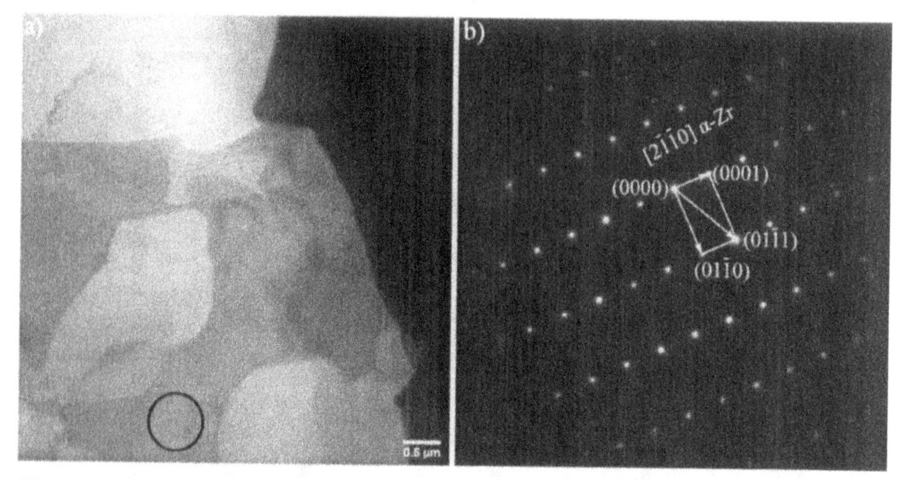

Figure 6 A continuous phase and its SAED pattern: a) TEM image showing the continuous phase, and b) the diffraction pattern form the circled area in a)

CONCLUSIONS

Phases in the matrix of the C_f/ZrC composites fabricated through RMI of a molten Zr-10at % Si alloy was studies by XRD, SEM/EDS, STEM/EDS, and TEM. The following conclusions are obtained.

(1) The XRD results indicate that there are four phases in the C_f/ZrC composites, i.e. Carbon, α-Zr, Zr_3Si and ZrC.
(2) Investigation by SEM/EDS and STEM/EDS indicates that it may have α-Zr, Zr_3Si and ZrC in the matrix of the C_f/ZrC composite.
(3) It was further confirmed by TEM that there exist α-Zr, Zr_3Si and ZrC in the matrix of the C_f/ZrC composite.

FOOTNOTES

*Member of The American Ceramic Society

ACKNOWLEDGEMENT

The authors would like to thank Dr. Noah Bodzin at Nanolab in UCLA for his help during sample preparation by using FIB.

REFERENCES

[1]Rajech, G; Bhagat, R B, Modelling micro-level volume expansion during reactive melt infiltration using non-isothermal unreacted-shrinking core models, Modelling and Simulation in Materials Science and Engineering, **6**, 771-786 (1998).
[2]William, Hillig, Melt Infiltration Approach to Ceramic Matrix Composites, J. Am. Ceram. Soc., **71**, C-96-C-99 (1988).
[3]Robert P. Messner and Yet-Ming Chiang, Liquid-Phase Reaction-Bonding of Silicon Carbide Using Alloyed Silicon-Molybdenum Melts, J. Am. Ceram. Soc., **73**, 1193-1200 (1990).

[4]Omprakash Chakrabarti and Probal Kumar Das, Reactive Infiltration of Si-Mo Alloyed Melt into Carbonaceous Preforms of Silicon Carbide, J. Am. Ceram. Soc., **83**, 1548-50 (2000).
[5]Yet-Ming Chiang, Robert P. Messner and Chrysanthe D. Terwilliger, Donald R. Behrendt, Reaction-formed Silicon Carbide, Materials Science and Engineering, **A144**, 63-74 (1991).
[6]Linhua Zou, Natalie Wali, Jenn-Ming Yang, Narottam P Bansal, and Dong Yang, Microstructural Characterization of a C_f /ZrC Composite Manufactured by Reactive Melt Infiltration, International Journal of Applied Ceramic Technology, **8**, 329-341 (2011).
[7]Linhua Zou, Natalie Wali, Jenn-Ming Yang, Narottam P Bansal, Microstructural Development of a C_f /ZrC Composite Manufactured by Reactive Melt Infiltration, Journal of European Ceramic Society, **30**, 1527-1535 (2010).
[8]H. Okamoto: Bull. Alloy Phase Diagrams, **11**, 513-19 (1990).
[9]Rossteutscher W., Schubert K.: "Strukturuntersuchungen in einigen T^{4-5}-B^{4-5}-Systemen", Z. Metallkd. **56**, 813–822 (1965).

Technologies for Innovative Surface Coatings

INTERACTIONS BETWEEN AMORPHOUS CARBON COATINGS AND ENGINE OIL ADDITIVES: PREDICTION OF THE FRICTION BEHAVIOR USING OPTIMIZED ARTIFICIAL NEURAL NETWORKS

Edgar Schulz[1,2], Thilo Breitsprecher[1], Yashar Musayev[2], Stephan Tremmel[1], Tim Hosenfeldt[2], Sandro Wartzack[1], Harald Meerkamm[1]
[1] Chair for Engineering Design of the Friedrich-Alexander University Erlangen-Nuremberg, Germany
[2] Schaeffler Technologies GmbH&Co. KG, Herzogenaurach, Germany

ABSTRACT

Amorphous carbon coatings are more and more used in combustion engines. In the valve train these coatings are applied in order to fulfill legislative guidelines concerning energy efficiency and CO_2 emissions. Up to now the effect of interactions between additives and such coatings on the friction is not sufficiently understood. Especially the high complexity of valve train systems and large experimental effort needed for a coating development show the need for a specific prediction of the friction behavior. Since an analytical prediction in such complex systems is not possible, always empirical studies are needed to determine the tribological behavior. This article presents the development and optimization of different multilayer artificial neural networks (ANNs) to predict the friction behavior on basis of tribological test data. For this a multitude of experiments were carried out by using various tribological test equipments whereby input parameters like type of coating, base oil, additives, temperature, pressure, etc. were varied systematically. The predictive capabilities of the ANN models were validated with experimental results. With a systematic variation of the learning rate and structure of the ANNs, a correlation coefficient from 0.69 up to 0.85 and relative absolute error of about 13% to 21% could be achieved.

INTRODUCTION

Today CO_2 reduction often is used as a synonym for friction reduction in the automotive industry. Above all, legislative guidelines, for example these of the European Union, to regulate CO_2 emissions and fleet consumption of automobile manufacturers are the driving forces for the improvement of energy efficiency and durability of modern gasoline engines.[1] This means, that it is getting more and more important to reduce friction and wear in highly stressed contacts of combustion engines. Therefore, the application of amorphous carbon coatings, used for example in the valve train, is increasing. In such tribological systems the counterpart, ground part, intermediate and environment interact one with another in a manifold manner (see Figure 1).[2] In the contact area between ground part and counter part elementary friction and wear mechanisms such as adhesion, abrasion, surface fatigue, deformation, elastic hysteresis and damping, tribochemical reactions etc. may occur. These mechanisms are overlapping in the temporal and spatial not detectable real material contact area. In addition highly doped lubricants, which are mainly used with highly stressed engine elements, exert influence on tribochemical reactions inside and outside of the tribological contact.[2]

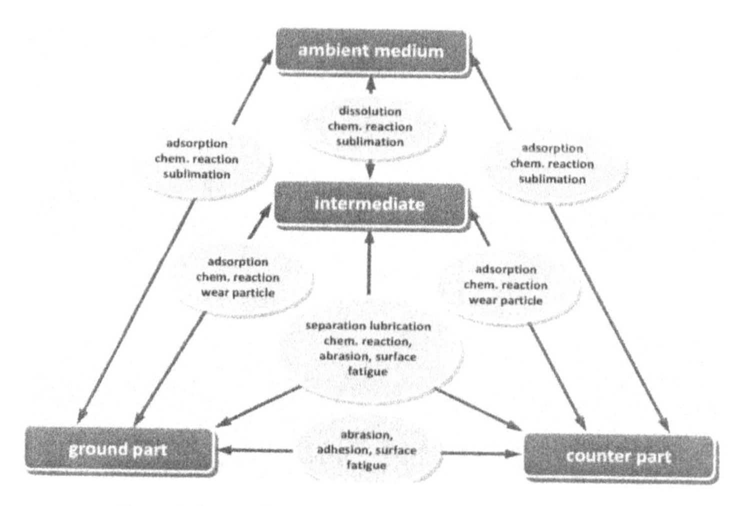

Figure 1: Interactions in a tribological system according to[2].

Due to these various reciprocal effects the behavior of such tribological systems becomes very complex, which means that the theoretical forecast of friction and wear is hardly possible at present. Therefore during development and/or optimization of each tribological system always experimental investigations mostly based on empirical knowledge are conducted. Thereby normally the focus is only on one machine part or even exclusively on the coating. Ideally such investigations are carried out on test rigs, which reproduce the real system as close as possible, for example on a driven cylinder-head test rig, to investigate the behavior of the machine part respectively the coating under real boundary conditions. However, such experiments are associated with a high temporal and financial effort. This often leads to the application of abstracted, cheap and time effective model tests or component tests like the ball-on-disk tribometer. However the transferability of these tests to the application is restricted. This indicates the need for a suitable method to predict the friction and wear behavior of tribological systems containing tribological coatings. With this development times and the amount of expensive tribological tests can be reduced. In the following this is going to be illustrated for the example of the tribological contact cam/coated bucket tappet.

THE BUCKET TAPPET VALVE TRAIN

The valve train is one of the main elements of combustion engines and is considerably involved in the friction losses of the combustion engine.[3] Figure 2 illustrates the portion of mechanical losses in an engine caused by the valve train depending on the revolutions per minute of the crank shaft.

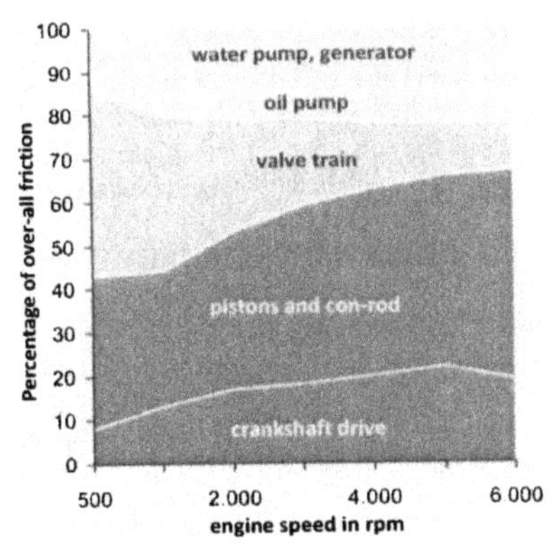

Figure 2: Mechanical losses of a gasoline engine with bucket tappet valve train.[4]

The valve train can take several forms depending on the position of the camshaft and design of the cam follower. Above all, the mechanical bucket tappet and the roller finger follower are at present the most common valve train systems.[5] With the roller finger follower the contact between cam and follower is designed as a rolling contact with low friction losses. However, the realization of a rolling contact is considerably more expensive.[3,6] This is as well reflected in the cost benefit ratio, as can be seen in Figure 3. Here the basis with 0% friction reduction and 100% costs is represented by the carbonitrided and tempered mechanical bucket tappet. A friction reduction of about 25% can be achieved by an optimized surface treatment of the standard bucket tapped. An additional reduction of friction down to the low level of a roller finger follower is possible by the application of a tailored coating.[6] With an adjustment of all components, that means for example the best coating for the respective lubricant or the respective cam and cam structure, it is possible to reduce the fiction level even more.

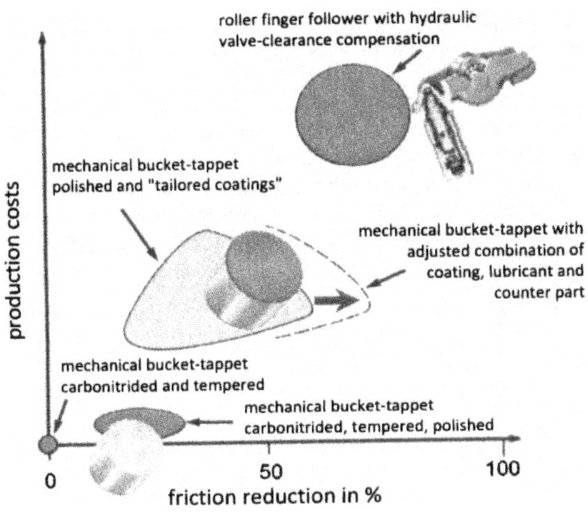

Figure 3: Cost-benefit ratio of different valve train modifications according to[6].

Considering the fact, that reduction of friction in the gasoline engine can be equated with a reduction of the CO_2-emission, legislative guidelines will increasingly influence this cost-benefit approach.[1] With this the greater expense for coating engine parts can be amortized by the reduction of CO_2 emission. As mentioned amorphous carbon coatings are more and more used in the valve train of gasoline engines. In the following these special coatings will be described more detailed.

AMORPHOUS CARBON COATINGS AND ITS INTERACTION WITH ADDITIVES CONTAINING LUBRICANTS

Since the mentioned interactions take place mainly at the surface of contacted bodies, friction and wear can be influenced by surface modification like texturing and coating. By coating the ground part for example, a further element is added to the structure of the tribological system (see Figure 4).[7] These surface modifications have to be adapted to the respective application and should be taken into account during the design phase of the tribological system, so the coating should be seen as a design element.

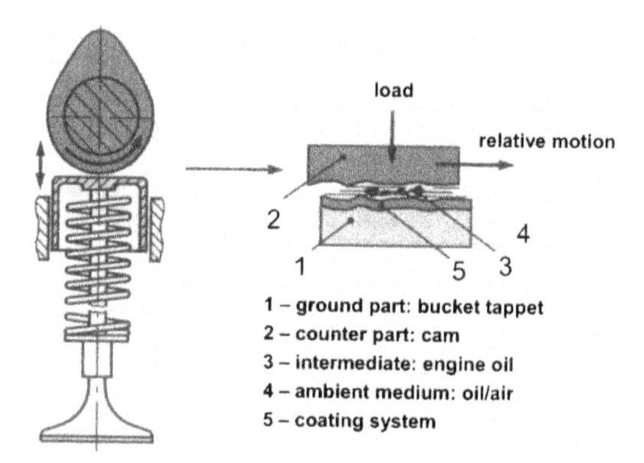

Figure 4: The five elements of a tribological system containing a coating

Amorphous carbon coatings are in particular suitable for this purpose, since its properties can be adjusted for different requirements. These characteristics are founded on its binding structure and hydrogen content. Amorphous carbon coatings consist of notably stable diamond like tetrahedral sp^3-bonds, which ensure a high wear resistance, and graphitic sp^2-bonds, that enable a slight slipping between the atomic hexagonal structured planes.[8,9] Regarding its structure and its properties amorphous carbon coatings can be located in a triangle between graphite, polymer and diamond (see Figure 5).

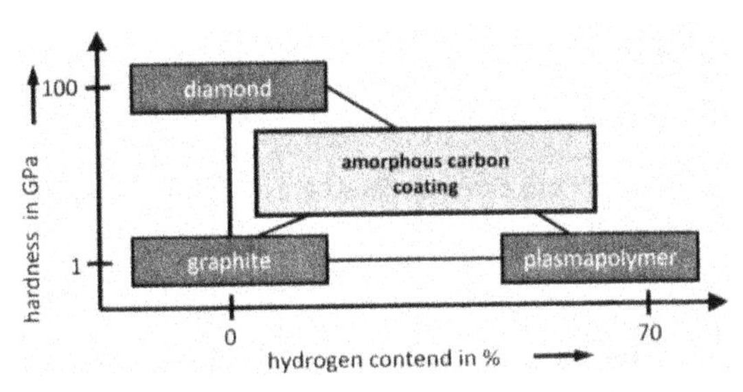

Figure5: Relationship between hydrogen contend an hardness of amorphous carbon coatings according to [8,9]

Depending on the manufacturing process, the coating parameters and the exact composition, it is possible to adjust the properties of the coatings up to the edges of the triangle. Additional the chemical structure and polarity of the coating and therefore the reactivity with surface active additives, can be influenced by doping the coating with metals (Cr, W, Ti, ...) or nonmetals (F, O, Mo, Si, ...).[8]
The advantages of amorphous carbon coatings under dry conditions are the low friction coefficient of about 0.1 to 0.2 in combination with high wear resistance.[9] In lubricated tribological systems working under mixed friction conditions the performance of amorphous carbon coatings depends on the interaction between all elements of the tribological system. Concerning its additive composition, actual engine oils are optimized for steel/steel contacts. Accordingly the chemical and physical behavior of additives, like anti-wear agents, extreme pressure agents and friction modifiers, in contact with steel is well known. If however engine parts are coated with amorphous carbon coatings, the chemical behavior of one element changes and therefore the behavior of the whole tribological system. The reason for this is that these additives (e.g. Mo-DTC, Zn-DTP, ...) cannot completely take effect on the inert and non-polar carbon surface. Through the changed reaction conditions even undesirable effects may occur leading to an increase of friction and wear.[10,11]

PREDICTION OF FRICTION AND WEAR

The goal during an application-based and market-oriented development of tribological coatings for reduction of CO_2 emissions is to provide a coated system solution respectively coated components with a maximum of customer benefit as quick and cost-efficient as possible. To reduce development time and cost coating, lubricant and counter part need to be adjusted purposefully to the particular application. This means for the system bucket tappet/ camshaft, that with a given component – for example camshaft, lubricant or coating – a recommendation for the design of the remaining components can be given.
One possibility to achieve this goal is the isolated determination of single effects, as published in [10-17]. Thereby the single effects are observed under fixed boundary conditions realized by abstraction and simplification of the tribological system in order to handle the complexity of the tribological processes. Thus in [10] the reactivity of the anti-wear agent Zn-DTP could be verified on an amorphous carbon coating containing hydrogen by the use of a ball-on-disk tribometer with rotating disc and an additional XPS analysis. However the adhesion strength of the resulting tribofilm was considerably weaker than on a steel surface. As a result the wear reducing effect of Zn-DTP cannot occur and on the contrary hard wear promoting particles are brought into the tribological contact by delamination of the reaction layer. Moreover, the formation of a reaction layer by Zn-DTP requires energy, which leads to an increase of the friction coefficient up to level of the steel/steel contact.[10] However it is possible to achieve a significant friction reduction in the range of fluid friction with hydrogen free carbon

coatings, when the lubricant is adjusted to the carbon surface. In [12] it was found, that special friction modifiers are especially effective with tetrahedral hydrogen free amorphous carbon coatings (ta-C). In a ball-on-disc tribometer with oscillating ball, ta-C coated disc and a GMO doped base oil friction coefficients below 0.02 were reached. This extreme low friction is explained by the hydroxylation of the ta-C surface leading to low shear forces in the interface between coating and lubricant.[12] With amorphous carbon coatings containing hydrogen such a drastic friction reduction was not reached yet. In [11] various amorphous carbon coatings were investigated with a lubricated block-on-ring test lubricated with a Mo-DTC friction modifier doped base oil. Here it was found, that Mo-DTC decomposes in the tribological contact under generation of MoS_2 and MoO_3. Although the friction coefficient was reduced with most amorphous carbon coatings, a rise in the wear amount was observed. MoO_3 seems to promote the oxidation of the amorphous carbon coatings.[11]

These described investigations were carried out with model tests, which can be assigned to the different categories of tribological test engineering as can be seen in Figure 6 on the left side. The aggregate tests and field tests are located further to the right in Figure 6. Here the transferability to the application is much better than on the left side, but it is to be expected that these test are more expensive and time-consuming.

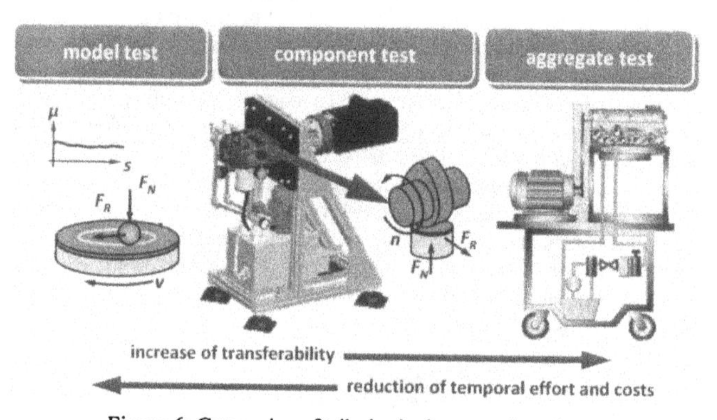

Figure 6: Categories of tribological test engineering

Just in such tribological systems like the contact between cam and coated bucket tappet with lubricants containing additives it is quite possible, that results out of different categories can be contradictory. This was proven in [18] by carrying out experiments with different amorphous carbon coatings on a ball-on-disc tribometer and a driven cylinder head rig, whereby no correlation between the results could be found. So based only on model tests it is not possible to fathom out all effects and superimpositions for a transfer to the real system cam/bucket tappet. Up to now this transferability can only be ensured when aggregate tests or even field tests are applied, whereby however the financial and temporal efforts become disproportionately high. But applying statistical methods with learning capabilities, the experimental effort can be reduced significantly. Thereby it is possible to determine expected output variables from given input variables. For example artificial neural networks allow a quite good prediction of friction and wear on the basis of a large amount of experimental data.

THEORY OF ARTIFICIAL NEURAL NETWORKS

Artificial neural networks (ANNs) try to simulate a biological neuronal system. For different tasks like pattern recognition, classification or regression, different kinds of ANN like the perceptron, kohonen net or multilayer ANNs exist. In this paper, different feed forward multilayer ANNs were built to compare its accuracy regarding the prediction of the friction behavior of tribological systems. Such networks comprise an input layer, one or more hidden layers and an output layer. In a feed-forward network, each layer is made up of neurons which are only interconnected via weighted links to the neurons in the next layer (see Figure 7).

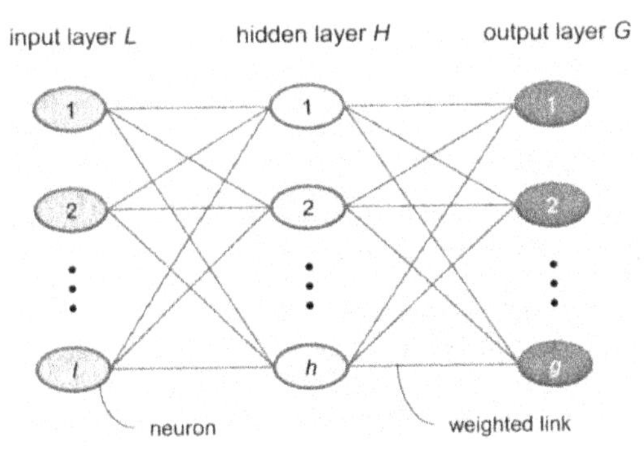

Figure 7: Schematic topology of a multilayer ANN

Functioning of a neuron

Hidden and output neurons represent the information-processing units of an ANN, while input neurons only pass on information namely the values of the input variables ($\chi_i = o_i$). Hereafter the functioning of a hidden neuron in the first hidden layer is explained based on neuron j (see Figure 8).The function of this neuron can be transferred to all neurons in further hidden layers as well as to neurons in the output layer.

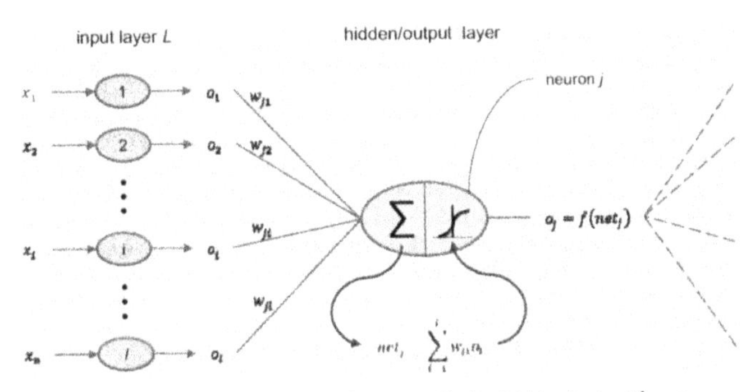

Figure 8: Functioning of neuron j in the hidden/output layer

The neuron j collects the outputs o_i from the neurons i in the previous layer with an integration or propagation function. This function reduces the incoming l arguments to a single value, called net input net_j generally, net_j will be calculated by taking the weighted sum of o_i:

$$net_j = \sum_{i=1}^{I} w_{ji} o_i \tag{1}$$

where w_{ji} represents the weight that determines the contribution of neuron i to the neuron j. Via net_j the actual activation state of the neuron j will be determined by the activation function $f(net_j)$. Finally, the actual activation state represents the output o_j of neuron j.

$$o_j = f(net_j)$$

where

$$f(net_j) = \frac{1}{1 + e^{net_j}} \tag{2}$$

In equation (2) the logistic or the sigmoid function is used for the activation function. Other possible functions are for example the linear or the sign function.[19, 20]

Training an artificial neural network via backpropagation

There are different algorithms for training an ANN. The well known and most famous training algorithm is the so called backpropagation algorithm (BPA). In the following, this algorithm will be explained on the basis of the network depicted in Figure**Error! Reference source not found.** 9.

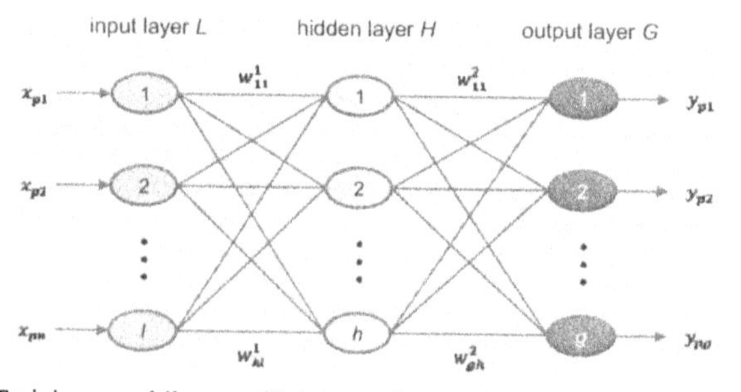

Figure 9: Training a multilayer artificial neural network via backpropagation algorithm

The training process via BPA will be carried out by learning on the basis of training examples. The training example p is a pair of the form (χ_{pi}, t_{pj}), where χ_{pi} are its input variables and t_{pj} are its output variables. During a learning process, χ_{pi} is fed to the ANN and the output of the network y_{pi} will be computed. Following, the error between t_{pj} and y_{pj} will be determined and the weights w_{ji} of the ANN will be adapted for the minimization of this error. For the determination of this error, BPA uses the total square error $E_p(w_{ji})$.

$$E_p(w_{ji}) = \frac{1}{2} \sum_{g \in G} (t_{pg} - y_{pg})^2 \tag{2}$$

The adaption of w_{ji} will be realized by the change of $\Delta_p w_{ji}$:

$$\Delta_p w_{ji} = \eta \frac{\partial E_p(w_{ji})}{\partial w_{ji}} \tag{3}$$

where η is the learning rate controlling the update step size.

For the deduction of the learning rule of BPA, equation (2) and equation (1) will be integrated to equation (3). Therefore, the neuron output y_j of the next lower layer will be recursively considered and so on. From the multiple application of the chain rule, the BPA-learning rule results in:

$$\Delta_p w_{ji} = \eta \delta_{pj} y_{pj} ,$$

with

$$\delta_{pj} = \begin{cases} y_{pj}(1 - y_{pj})(t_{pj} - y_{pj}) & \text{if } j \text{ is output neuron} \\ y_{pj}(1 - y_{pj}) \sum_g \delta_{pg} w_{gj} & \text{if } j \text{ is hidden neuron} \end{cases} \quad (4)$$

This learning rule determines the change of w_{ji} from neuron i to neuron j. For the output neurons, the factor $(t_{pj} - y_{pj})$ provides an adaption of w_{ji} proportional to $E_p(w_{ji})$. For the hidden neurons, δ_{pj} of the neuron j will be calculated recursively by all changes of δ_{pk} of the next higher layer [19, 20].

PREDICTION OF THE FRICTION BEHAVIOR OF TRIBOLOGICAL SYSTEMS USING ARTIFICIAL NEURAL NETWORKS

Within this paper, different ANNs were modeled to investigate its suitability for the prediction of the friction behavior of tribological systems. In the following, based on the description of the data, the training of ANNs and its application on new data will be described.

Data Collection

Precondition for a predictive modeling with ANNs is the availability of a sufficient large amount of empirical test data. For this database already existing test data from a driven cylinder head test rig (Figure 6 right) were pulled up. Furthermore, a multitude of experiments was carried out by a systematic variation of the explanatory variables with a ball-on-disk tribometer (Figure 6 left). As already mentioned, there is an obvious discrepancy between the results of these two tests. Therefore, a novel valve train test rig, which should close the gap between the model test and the aggregate test, was used. In Figure 6 (middle) this test rig is illustrated. It is designed as a single valve test rig and reproduces the system cam/bucket tappet as realistic as possible. Thereby a genuine camshaft section with the associated bucket tappet and valve spring is used. This valve train model test rig provides the opportunity to determine the friction and normal forces appearing in the tribological contact separately. The used database generated with all these different tribological methods includes a total amount of 262 experiments (N_{DO}).

In case of the tribological contact between cam and coated bucket tappet, the output variable is the friction coefficient. The input variables are the type of coating its hardness, Young's modulus and the surface roughness. Thereby beside uncoated bucket tappets and a nitride chromium hard coating, six different amorphous carbon coatings were investigated. The used amorphous carbon coatings differ in its hydrogen contend and its doping elements. The lubricant additives, its concentration, the base oil and its viscosity were added to the database as well. Here, PAO and mineral base oils as well as the additives Mo-DTC, Zn-DTP and GMO were taken into account. Furthermore, the stress parameters of the tribological experiment were varied as input variables, so relative velocity, oil temperature and contact pressure. Table 1 shows the meta data view about the input variables, the output variable, the defined variation, relative measurement errors and correlation to the friction coefficient.

Table 1: Input and output variables for the predictive modeling with an ANN

category	name	description	unit	range	relative error
coating parameters	input 1	type of coating	-	[0 ; 7]	-
	input 2	doping	-	[0; 2]	-
	input 3	bonding structure	-	[0 ; 3]	-
	input 4	hardness	HV	[875 ; 5500]	10 %
	input 5	Young's modulus	GPa	[110 ; 290]	10 %
	input 6	roughness	µm	[0.005 ; 0.467]	2 %
oil parameters	input 7	base oil	-	[0 ; 2]	-
	input 8	viscosity	mm^2/s	[30 ; 50]	3 %
	input 9	additive element concentration	mg/Kg	[0 ; 1000]	7 %
	input 10		mg/Kg	[0 ; 1000]	
	input 11		mg/Kg	[0 ; 760]	
	input 12		mg/Kg	[0 ; 2400]	
	input 13		mg/Kg	[0 ; 2400]	
	input 14		mg/Kg	[0 ; 10000]	
tribological parameters	input 15	tribo test	-	[0; 2]	-
	input 16	temperature	°C	[22 ; 120]	5 %
	input 17	pressure	MPa	[200 ; 2940]	1 %
	input 18	velocity	m/s	[0.01 ; 3.5]	1 %
friction	output	µ	-	[0.019 ; 0.171]	5 %

The repetitive accuracy of the friction measurements was determined by repeating a measurement several times at the same conditions. These results show that the relative variance of these measurements is in a range of about 5% of the friction value.

TRAINING ARTIFICIAL NEURAL NETWORKS

The objective of training an ANN is its adaption to a given problem. The performance of ANNs depends on several factors like the database size, data preprocessing, training settings and the topology of the networks. Within this paper, the effect of the learning rate, the number of hidden layers as well as the number of hidden neurons was examined regarding the prediction of the friction behavior of tribological systems. For the training and the application of multilayer networks, the data mining software RapidMiner™ was used. During the parameter studies only one parameter was varied, the other parameters were kept on default value. The default settings of ANNs in RapidMiner™ regarding the parameters to be varied are:

- Learning rate = 0,3
- Number of hidden layers = 1
- Number of hidden neurons = [("number of input neurons" + "number of output neurons")/2]+1

Partitioning of data

The development of a multilayer ANN requires the partition of the given dataset into a training set and a test set. The training set is used for building an ANN, whereas the test set is used for its performance estimation. For splitting a dataset, different methods like the holdout method, bootstrap and k-fold cross validation exist. The holdout method and the bootstrap approach subdivide a given data set into two disjoint sets by a random sampling. The difference between both methods is that the

bootstrap approach bases on random sampling with replacement and the holdout method does not. The k-fold cross validation splits the dataset randomly into k approximately equal disjoint subsets. The following learning procedure is executed k-times in which per procedure one subset is used for testing and the remainder is used for training. Finally, the resulting k error estimates are summarized and averaged to the total error estimate. According to [21] the recommended value for k is 10. For a more reliable error estimate, a stratified k-fold cross-validation can be applied to reduce the effect of the random partition of the dataset. [21, 22, 23]
According to [21] the standard estimation technique for a small dataset is the stratified k-fold cross-validation. This was also confirmed from [23] by the study of cross-validation and bootstrap for accuracy estimation and model selection. Therefore, within this paper a 10 times 10-fold cross-validation was used for the performance estimation of the various multilayer ANNs.

Performance measures for the assessment of artificial neural networks

In general, a couple of performance measures can be used for the assessment of regression models. In this paper, the applied measures and its underlying formulas are depicted in Table 2. The characteristic of root mean-squared error (RSME) compared to the mean absolute error (MAE) is that this tends exaggerate the effect of outliers, whereas the mean absolute error does not do this. In some cases, the assessment via preceding measure is not strong enough, thus a relative measure like the relative absolute error (RAE) has to be taken into account. For the estimate of the statistical correlation between the real target variable t_j and the predicted target variable y_j, the correlation coefficient can be applied. This coefficient ranges from "+1" through "0" to "-1". A coefficient value of "+1" means a perfect correlation, whereas a coefficient value of "0" means no correlation. A perfect negative correlation is characterized by a coefficient value of "-1".

Table 2: Performance measures for regression techniques [21]

Performance measure	Formula
Root mean-squared error (RSME)	$\sqrt{\frac{\sum_{j=1}^{n}(t_j - y_j)^2}{n}}$
Mean absolute error (MAE)	$\frac{\sum_{j=1}^{n}\lvert t_j - y_j\rvert}{n}$
Relative absolute error (RAE)	$\frac{1}{n}\sum_{j=1}^{n}\frac{\lvert t_j - y_j\rvert}{y_j}$
Correlation coefficient (CC)	$\frac{S_{ty}}{\sqrt{S_t S_y}}$, where $S_{ty} = \frac{\sum_j (t_j - \bar{t})(y_j - \bar{y})}{n-1}$, $S_t = \frac{\sum_j (t_j - \bar{t})^2}{n-1}$, $S_y = \frac{\sum_j (y_j - \bar{y})^2}{n-1}$

VARIATION OF SETTING PARAMETERS

Learning rate

The learning rate η determines the degree to which the weights w_{ji} are changed at each training step. A high learning rate indicates on the one hand a fast adaption of w_{ji} and on the other hand a high possibility of oscillation about the optimal solution. A small learning rate, however, leads to a low change of w_{ji} at each training step, but increases the possibility to reach the optimal solution. In literature, there are different rules of thumb to define the learning rate.[24] suggests $\eta = 0.3 - 0.6$, [25] recommends $\eta = 0.0 - 1.0$ and [26] suggests $\eta = 0.01 - 0.9$.

Within this article, the learning rate for training an ANN was varied in the range from 0,1 to 0,4. A higher learning rate could not be used because RapidMiner™ stopped the training. As a result of this parameter study, the theory, that the decrease of the learning rate leads to a better prediction accuracy, could be verified by the performance measures of the modeled ANNs depicted in Table 3. After this, a decrease of the learning rate causes on the one hand the decrease of all error measures and on the other hand the increase of the correlation coefficient. An exception, however, represents the ANN_3 because its correlation coefficient is smaller than the one of ANN_4.

Table 3 Performance of ANNs depending on learning rate η

	η	RSME	MAE	RAE	CC
ANN_1	0,1	0,014	0,011	13,75%	0,852
ANN_2	0,2	0,016	0,012	14,67%	0,814
ANN_3 (ANN_def)	0,3	0,018	0,013	15,96%	0,765
ANN_4	0,4	0,019	0,014	17,58%	0,788

Number of hidden layers

The number of hidden layers depends on the complexity of the problem to be solved. Linear problems can be solved by ANNs with no hidden layers, while nonlinear problems should be treated by ANNs with at least one hidden layer [25]. [26] recommends the use of two hidden layers, whereas [27] suggests the use of only one hidden layer because two or more hidden layers do not improve the network performance substantially but increase the training time.

The influence of the number of hidden layers on the performance of ANNs was examined by designing networks with 1-, 2-, 3- and 4- hidden layers (see Table 4). For the given database, the recommendation of [26] to use an ANNs with 2-hidden layers could be confirmed. Furthermore, it can be stated, that ANNs with more than 2-hidden layers compared to an ANN with 1-hidden layer do not show always smaller error measures and higher correlation coefficients.

Table 4 Performance of ANNs depending on number of hidden layers (NHL)

	NHL	RSME	MAE	RAE	CC
ANN_5 (ANN_def)	1	0,018	0,013	15,96%	0,765
ANN_6	2	0,016	0,012	15,51%	0,807
ANN_7	3	0,017	0,013	16,46%	0,815
ANN_8	4	0,017	0,014	17,45%	0,789

Number of hidden neurons

The determination of the appropriate number of hidden neurons (NHN) is not straightforward and represents one of the most critical challenges in the modeling of ANNs. The choice of the quantity of hidden neurons has a massive effect on the training time and on the overfitting or underfitting of the networks to the training set, respectively. According to [26] the optimal number of hidden neurons has to be found by trial-and-error. For a starting point, following heuristics can be applied: In [28] NHN is equivalent to 50 % of the quantity of input and output neurons, while in [29] NHN corresponds to 75 % of the quantity of input neurons. [30], however, calculates NHN by $N_{DO}/[R\cdot(N_{inp}+N_{out})]$, where N_{DO} are the number of data objects in the database, N_{inp} is number of input neurons, N_{out} is number of output neurons and R ranges between 5 and 10 (see Table 5).

Table 5 Number of hidden neurons for the given database

	Rev[29]	Rev[28]	Rev[30]						Rapid-Miner
			R						
			5	6	7	8	9	10	
calculated NHN	13,5	9,5	2,75	2,29	1,96	1,72	1,53	1,37	10
applied NHN	13	9	3	2				1	10

For investigation the effect of the number of hidden neurons on the performance of ANNs, the previous mentioned heuristics for the determination of NHN were applied considering the given database (N_{inp}=17, N_{out}=1, N_{DO}=262).

According to Table 6, the ANN_12 with 9 hidden neurons is the best network according to all four metrics because it has the smallest value for each error measure and the largest correlation coefficient. An increase as well as a decrease of NHN leads to an increase of the error measure and to a decrease of the correlation coefficient.

Table 6. Performance of ANNs depending on the number of hidden neurons (NHN)

	NHN	RSME	MAE	RAE	CC
ANN_9	1	0,020	0,016	20,90%	0,689
ANN_10	2	0,019	0,015	17,98%	0,748
ANN_11	3	0,019	0,015	18,42%	0,764
ANN_12	9	0,017	0,012	16,14%	0,822
ANN_13 (ANN_def)	10	0,018	0,013	16,31%	0,775
ANN_14	13	0,019	0,015	18,97%	0,756

DEPLOYMENT OF ANNS ON NEW DATA

For the prediction of friction coefficients, the default ANN of RapidMiner™(ANN_def) and the best networks of each parameter study (ANN_1, ANN_6, ANN_12) were applied on the input variables of 5 new instances, which were not a part of the training data set. In Figure 10, the corresponding prediction results of each ANN are compared with the real friction coefficient. Eye-catching is that the predicted friction coefficients and the real friction coefficients match sufficiently well. Merely, the network ANN_1 possesses difficulties regarding the prediction of the friction coefficient for test data 3. The reason for this could be attributed to the overfitting of this network because overfitting can be accompanied by small learning rate.

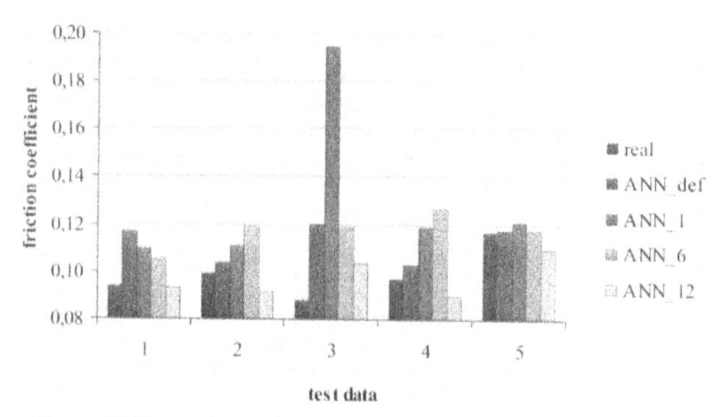

Figure 10 Comparison of the real and predicted friction coefficient

Furthermore, the bar chart in Figure 10 indicates, that the networks ANN_def as well as ANN_12 are capable to predict the friction coefficients very well in which its prediction quality varies from test data to test data differently strong. In accordance with Table 7, the smallest prediction error regarding all three error measures possesses the ANN_12 with its adjusted number of hidden neurons. The second best ANN represents the default network of RapidMiner™ (ANN_def) which differs from the ANN_12 by just one additional hidden neuron. The ANN_6 with its two hidden layers and its 10 hidden neurons per each hidden layer turns out as the third best network. According to all three error measures, the worst prediction model is the ANN_1.

Table 7. Performance of ANNs applied on new data

	RSME	MAE	RAE
ANN_def	0,018	0,013	14,58%
ANN_1	0,050	0,032	35,37%
ANN_6	0,022	0,019	20,20%
ANN_12	0,009	0,008	7,90%

CONCLUSION AND OUTLOOK

In order to reduce the temporal and financial effort for tribological experiments, different artificial neural networks were used to predict the friction behavior of lubricated tribological systems containing amorphous carbon coatings. Precondition for the modeling of artificial neural networks is the availability of a sufficient large database which contains case studies for the training of such networks. Therefore, 262 different experiments were carried out by variation of coating, lubricant, lubricant additives and stress collective.

The prediction accuracy of artificial neural networks depends on several factors like the quality of the raw data and the training process. Within this article, the effect of setting parameters in the training process on the performance of artificial neural networks was examined. It was shown, that the learning rate, the number of hidden layers and the number of hidden neurons affect the performance of an ANN. With a systematic variation of these three parameters, a correlation coefficient from 0.69 up to 0.86 and relative absolute error of about 13% to 21% could be achieved. Finally, the best ANN of each

parameter study were applied on new data in which the network with the adjusted number of hidden neurons turned out as the best network according to all performance measures (e.g. relative absolute error of 8 %).
In summary, the ANNs are able to predict the friction coefficient for different combinations of coating, lubricant or lubricant additives very well. For further improvements of the performance of ANNs regarding the prediction of friction coefficient, optimization algorithms like the evolutionary algorithm can be used to optimize networks to the respective database. Furthermore, the prediction accuracy of ANNs can be increased by preparation of the raw data. Applied methods here are for example data transformation or data reduction.

ACKNOWLEDGEMENT

This work was supported by the German Research Foundation (DFG) within the scope of the Transregional Collaborative Research Centre on Sheet-Bulk Metal Forming (SFB/TR 73) and by the Federal Ministry of Economics and Technology (BMWi) within the scope of the PEGASUS project (Progressive Benefit of Energy Efficiency in the Power Train through Coating Materials and Lubricants).

REFERENCES

[1] Regulation (EC) Nr. 443/2009 of the European Parliament and of the Council of 23 April 2009
[2] H. Czichos, K.-H. Habig: Tribologie Handbuch, Wiesbaden: Vieweg, 2010.
[3] P. Gutzmer, MTZ, 68 (2007), 243
[4] F. Koch, U. Geiger, tribological symposium of GfT and DGMK, Göttingen, 05.-06.11.1996.
[5] M. Lechner, R. Kirschner, in: R. van Basshuysen, F. Schäfer (Eds.): Handbuch Verbrennungsmotor. Grundlagen, Komponenten, Systeme, Perspektiven. Braunschweig/Wiesbaden: Vieweg, 2002.
[6] A. Ihlemann, G. Eggerath, T. Hosenfeldt, U. Geiger, 14th Aachener colloquium vehicle and motor technology, Aachen, 04.-06.10.2005.
[7] Y. Musayev, Verbesserung der tribologischen Eigenschaften von Stahl/Stahl-Gleitpaarungen für Präzisionsbauteile durch Diffusionschromierung im Vakuum, 2001.
[8] VDI guidline 2840: Kohlenstoffschichten – Grundlagen, Schichttypen und Eigenschaften. Düsseldorf: VDI, 2005.
[9] J. Robertson, Material Science and Engineering R37 (2002) 129-281.
[10] S. Equey, S. Roos, U. Mueller, R. Hauert, N. D. Spencer, R. Crockett, Wear 264 (2008) 316–321.
[11] T. Shinyoshi, Y. Fuwa, Y. Ozaki, Spring Academic Lectures of the Automobile Technology (2007)
[12] M. Kano, Tribology Letters, Vol. 18, No. 2, (2005) 245-251
[13] A. Erdemir, Tribology International 37 (2004) 1005–1012
[14] T. Haque, A. Morina, A. Neville Surface & Coatings Technology 204 (2010) 4001–4011
[15] T. Haquea, A. Morina, A. Neville , R. Kapadiab, S. Arrowsmith Wear 266 (2009) 147–157
[16] Ksenija Topolovec-Miklozic, Frances Lockwood , Hugh Spikes Wear 265 (2008) 1893–1901
[17] B. Podgornik, J. Vizintin, Surface & Coatings Technology 200 (2005) 1982 – 1989
[18] E. Schulz, Y. Musayev, S. Tremmel, T. Hosenfeldt, S. Wartzack, H. Meerkamm, Magazin für Oberflächentechnik 65 (2011) Nr. 1-2, S. 18-21.
[19] D. E. Rumelhart, G. E. Hinton, R. J. Williams,: Learning internal representation by error propagation. In: Rumelhart, D. E., McCllealand, J. L. (Eds.): Parallel Distributed Processing – Exploration in the Micorstructure of Cognition, Vol. 1. MIT press, Cambridge. 1986
[20] S. Haykin, Neural Networks – a Comprehensive Foundation, New Jersey: Prentice-Hall, 2nd edition, 1999
[21] I. H. Witten and F. Eibe: Data Mining – Practical Machine Learning Tools and Techniques, 2005 (Morgan Kaufmann, San Francisco).
[22] P. Tan, M. Steinbach and V. Kumar, Introduction to Data Mining, 2006 (Pearson Education, Boston).
[23] R. Kohavi, A study of cross-validation and bootstrap for accuracy estimation and model selection, in Proc. Proceedings of the Fourteenth International on Artificial Intelligence, IJCAI, , 2010, pp. 1137-1143
[24] J. Zupan, J. Gasteiger, Anal.Chim.Acta 248, 1-30, 1991
[25] L. Fu: Neural networks in Computer Intelligence. McGraw-Hill, New York, 1995
[26] D. Kriesel: A Brief Introduction to Neural Networks. 2005 http://www.dkriesel.com/science/neural_networks
[27] R. J. Abrahart, L. See, Hydrological Processes 14, pp. 2157-2172
[28] S. Piramuthu, M. Shaw, J. Gentry, Decision Support Systems 11 (5) (1994) p. 509 -525
[29] B. A. Jain, B. R. Nag, Decision Sciences 26 (3) (1995) p. 283-302
[30] M. N. Jadid, D. R. Fairbairn, Enggn Appl. Artif. Intell. Vol 9 No. 3. Pp. 309-319. 1996

CONTROL OF OXYGEN CONTENT WITH OXYGEN GAS INTRODUCTION IN Cr(N,O) THIN FILMS PREPARED BY PULSED LASER DEPOSITION

K. Suzuki, T. Endo, T. Fukushima, A. Sato, T. Suzuki, T. Nakayama, H. Suematsu and K. Niihara
Extreme Energy-Density Research Institute, Nagaoka University of Technology, Nagaoka, Japan

ABSTRACT

Hard coatings, such as CrN, are used in cutting tools. By partial replacement of the N in CrN with O, Cr(N,O) thin films with a B1 (NaCl type) structure can be produced. Furthermore, the hardness of these thin films can be increased by increasing the oxygen content. In our previous work, thin films were prepared by depositing Cr vapor in N_2 or NH_3 ambient gas with residual oxygen, but controlling the oxygen content precisely was difficult. In this work, O_2 gas was introduced with precise control of oxygen content in Cr(N,O) thin films prepared by pulsed laser deposition. After the chamber was evacuated to a pressure of 2.5×10^{-5} Pa, O_2 gas was introduced and nitrogen plasma from a RF radical source was applied. They were then characterized by X-ray diffraction, infrared spectroscopy and electron energy loss spectroscopy. It was found that the oxygen content of the thin films changed from 0 to 62 mol % with increasing oxygen partial pressure (P_{O2}). The thin films with only the B1-Cr(N,O) phase were prepared under $P_{O2} < 7.5\times10^{-5}$ Pa. The oxygen content of the Cr(N,O) thin films was successfully controlled by P_{O2}.

INTRODUCTION

Hard coatings are applied on cutting tools to improve their cutting characteristics. Such coating materials with good performances in high-speed processes, good cost saving and faster dry process are highly favourable. Diamond-like carbon (DLC), cubic boron nitride (c-BN) and transition metal nitride have been widely used as hard coating materials. In particular, chromium nitride (CrN) has excellent properties such as oxidation resistivity and chemical stability[1, 2]. However, CrN thin films typically have low hardness compared to other materials such as TiN and TiAlN, limiting the use of CrN coatings.

In our previous work, chromium oxynitride (Cr(N,O)) thin films were prepared by pulsed laser deposition (PLD)[3]. The Cr(N,O) had a B1 (NaCl-type) structure similar to the Cr(N,O) prepared by the partial replacement of N in CrN with O. The hardness of Cr(N,O) thin films increased with increasing oxygen content (x) and the maximum value exceeded 30 GPa. It is entirely possible that the cause for this is solution hardening. However, this has not been clarified. Materials with an added fourth element to Cr(N,O) for improving hardness have also been studied[4, 5]. For example, chromium magnesium oxynitride ((Cr,Mg)(N,O)) thin films have a high hardness of more than 35 GPa[4].

Oxygen content control is required for identification of the solubility limit and elucidation of high hardness. Until now, Cr(N,O) thin films were prepared by depositing Cr vapor in N_2 or NH_3 ambient gas with residual oxygen. However, precise oxygen content control was difficult in this method. In the current work, an intuitive method using modification of the oxygen partial pressure (P_{O2}) for precise oxygen content control is suggested. A highly reactive atmosphere (pure O_2 and N radical) in new chamber is used in this method. Our goal in this study is to investigate the highly precise control of oxygen content by varying P_{O2} .

EXPERIMENTAL

Figure 1 shows a schematic illustration of the apparatus used for preparing the thin films. The ablation plasma was produced by irradiating an Nd: yttrium aluminum garnet laser (355 nm) onto a Cr target (99.9% purity). The laser was electro-optically Q-switched by a Pockels cell to produce intense pulses of short duration (7 ns). The deposition time was 5 h at a laser-pulse repetition rate of 10 Hz. The deposition surface area was 1 cm^2 on a single-crystal (100)-oriented silicon substrate placed at a distance of 45 mm from the target. The film thickness was approximately 100 nm on this condition. The substrate temperature was controlled at 973 K using an infrared lamp heater.

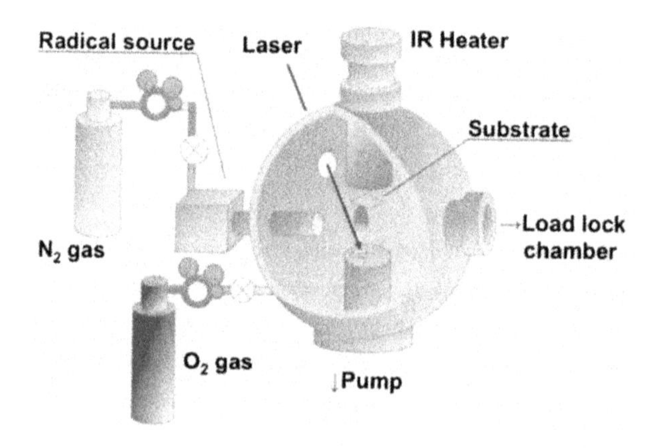

Figure 1. Schematic illustration of the apparatus used for preparing thin films.

The chamber was first evacuated to a pressure of 2.5×10^{-5} Pa using a rotary pump and a turbo molecular pump, and the chamber was then filled with oxygen gas (>99.99995 vol.% purity) and nitrogen plasma (>99.99995 vol.% purity) from an RF radical source. Both continuous pumping and introduction of O_2 gas and N plasma were carried out during the deposition. In order to change the oxygen content of the thin films, P_{O2} was varied. The thin films were prepared under a fixed total pressure (1.5×10^{-2} Pa).

The compositions of the thin films were determined by Rutherford backscattering spectroscopy (RBS) and electron energy loss spectroscopy (EELS). The crystal structures of the thin films were studied by X-ray diffraction (XRD) using CuKα radiation (0.154 nm). The chemical bonding states were estimated by Fourier transform infrared spectroscopy (FT-IR). The hardness of the thin films was measured by nano-indentation testing under a load of 0.07 mN using a Berkovich indenter. The microstructures of the thin films were observed using a field emission transmission electron microscope (FE-TEM) with a 200 kV acceleration voltage. The TEM samples were made by scratching the thin films with a diamond pen.

RESULTS AND DISCUSSION

Oxygen content

Figure 2 shows the oxygen content of the thin films as a function of P_{O2}. From the results of RBS and EELS measurements, it was found that the thin films contained chromium, nitrogen and oxygen. The Cr/(N+O) and O/N ratios were determined by RBS and EELS, respectively. As P_{O2} increased, x increased linearly from 0 to 62 mol%. The oxygen content of the Cr(N,O) thin film was thus successfully controlled by appropriately adjusting P_{O2}.

Figure 3 shows the composition of the thin films. When there was no oxygen in the thin films, the chromium content was about 50 mol%. In contrast, the chromium content of the thin films became closer to that of Cr_2O_3 when the oxygen content was increased. As we will describe later, the thin films have a B1 structure. Therefore, this indicated that vacancy was formed in the Cr site by the replacement of N with O.

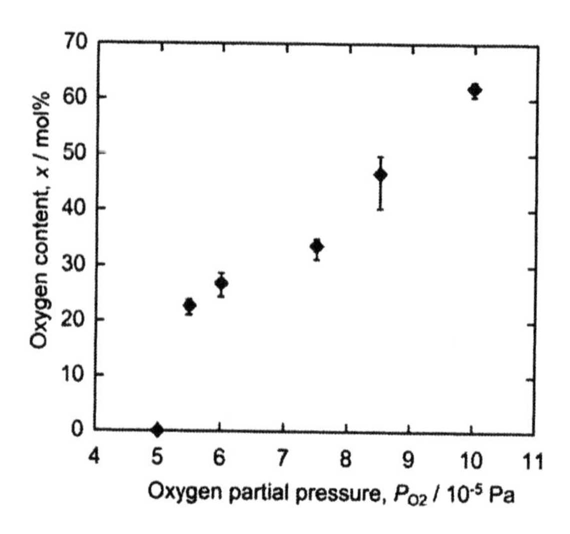

Figure 2. The oxygen content in the thin films as a function of oxygen partial pressure.

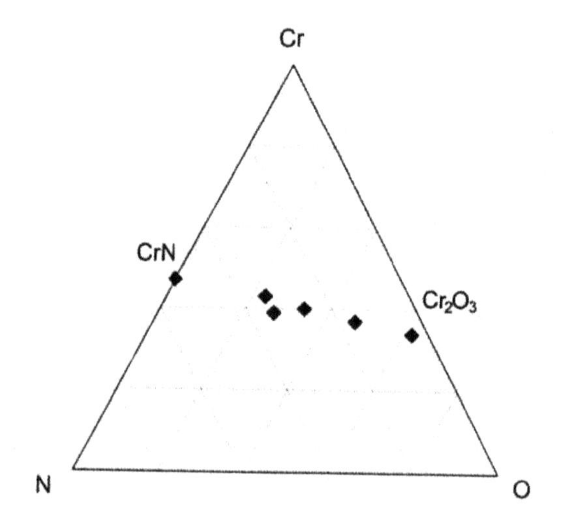

Figure 3. The composition of the thin films

Phase identification

Figure 4 shows XRD patterns of the thin films. Peak positions and relative intensities for CrN and Cr_2O_3 in International Centre for Diffraction Data (ICDD) are also included for comparison. It was found that all samples included a B1 phase based on CrN. In addition, only the sample formed with P_{O2}= 10×10^{-5} Pa included a Cr_2O_3 phase. The peaks due to the B1 structure became broad with increasing P_{O2}. Figure 5 shows the lattice constants of the B1 phase of each thin film calculated from the diffraction peaks. The lattice constant decreased with increasing P_{O2}. The change in the lattice constant indicated solution of oxygen in the B1 phase.

Figure 6 shows the FT-IR spectra of the thin films. The absorption spectra of the samples formed with P_{O2} less than or equal to 7.5×10^{-5} Pa showed a broad peak mainly at 500 cm^{-1} due to the Cr-N bond. Absorption peaks due to Cr_2O_3 were not observed. On the other hand, the absorption spectra of the samples formed with P_{O2} greater than or equal to 8.5×10^{-5} Pa show peaks due to the Cr_2O_3 bond. From the results in Figures 4 and 6, it was found that the thin films with only the B1-Cr(N,O) phase were prepared under P_{O2} less than or equal to 7.5×10^{-5} Pa.

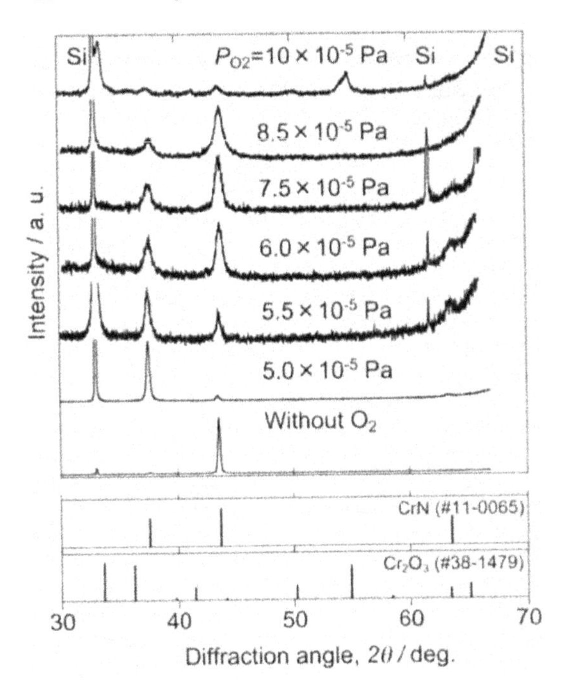

Figure 4. XRD patterns of the thin films prepared at different oxygen partial pressures.

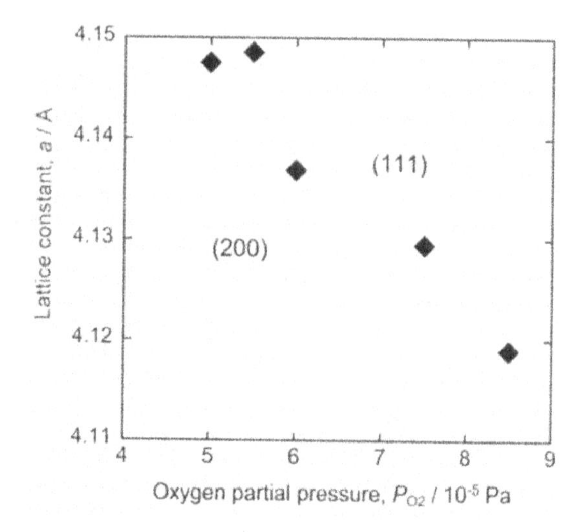

Figure 5. The lattice constants of thin films calculated from the XRD patterns. The solid and hollow symbols indicate the lattice constants calculated from the diffraction of 111 and 200. These values were calibrated from the diffraction of Si substrate.

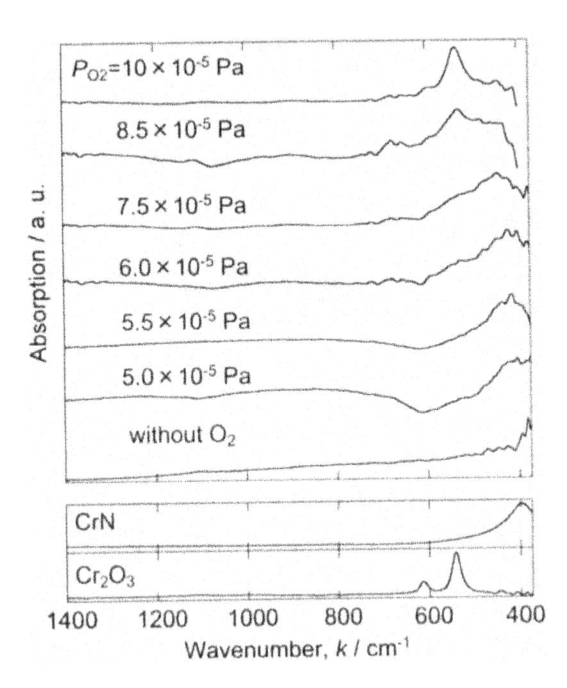

Figure 6. FT-IR spectra of the thin films prepared at different oxygen partial pressures. Reference data for CrN and Cr_2O_3 are also included[6].

Microstructure and Hardness

Figure 7 shows the bright field images (BFI), dark field images (DFI) and selected area diffraction patterns (SAD). The DFI were taken from the 200 diffraction of the SAD. It was found that grain size of the thin films was approximately 100 nm and did not change with increasing P_{O2}. It is entirely possible that the cause for this is fixed total pressure. Figure 8 shows the indentation hardness data for the thin films as a function of P_{O2}. The results of phase identification by XRD and FT-IR are also shown at the top of Figure 8. The hardness of the thin films increased with increasing P_{O2} up to 7.5×10^{-5} Pa. Above 7.5×10^{-5} Pa, the hardness decreased. The thin films showed a maximum hardness value of 32 GPa. The maximum value was found around the solubility limit. Because the grain size did not change, it was found that high hardness was not achieved by the Hall-Petch relationship.

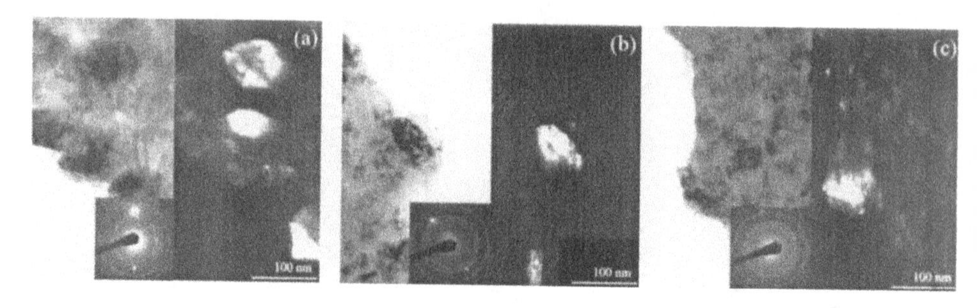

Figure 7. Bright field (left) and dark field (right) under excitation of the fcc (200) reflection of the thin films prepared at (a) P_{O2}= 5.0×10^{-5} Pa, (b) P_{O2}= 7.5×10^{-5} Pa and (c) P_{O2}= 8.5×10^{-5} Pa.

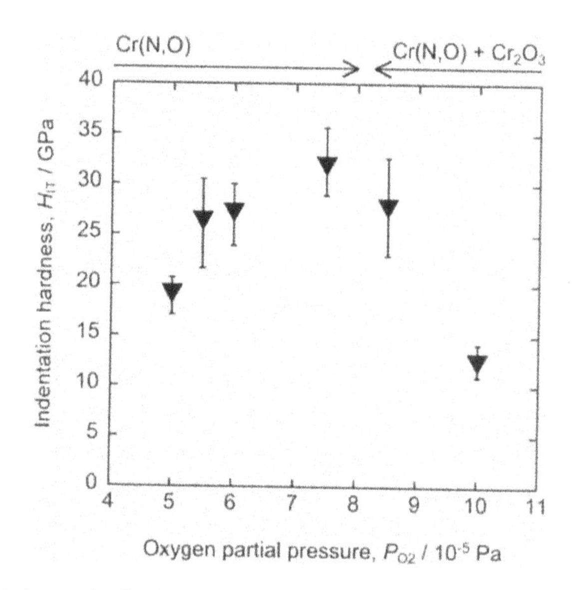

Figure 8 Nano-indentation hardness of the thin films as a function of oxygen partial pressure.

CONCLUSION

From the above results, it was found that the oxygen content in the Cr(N,O) thin films was successfully controlled by varying P_{O2}. The thin films had oxygen content up to 62 mol%. It was found that the crystal structure of all the samples have a B1 structure. In the case of high P_{O2} ($\geq 8.5\times10^{-5}$ Pa), Cr_2O_3 exists as a second phase. From TEM observations, it was found that grain size of the thin films did not change with increasing P_{O2}. The hardness of the thin films increased with increasing P_{O2}; however, it decreased for $P_{O2}\geq 8.5\times10^{-5}$ Pa. The thin films showed a maximum hardness value of 32 GPa around the solubility limit. Because the grain size did not change, it was found that high hardness was not achieved by the Hall-Petch relationship.

ACKNOWLEDGEMENT

This work was supported by Grant-in-Aid for Scientific Research 22686069.

REFERENCES

[1]T. Hurkmans, D.B. Lewis, H. Partiong, J.S. Brooks, and W.D. Müntz: Influence of ion bombardment on structure and properties of unbalanced magnetron grown CrN_x coatings, *Surface & Coatings Technology*, **114**, 52-59 (1999).

[2]J.-N. Tu, J.-G. Duh, and S.-Y. Tsai: Morphology, Mechanical properties, and oxidation behavior of reactively sputtered Cr-N films, *Surface & Coatings Technology*, **133**, 181-185 (2000).

[3]J. Inoue, H. Saito, M. Hirai, T. Suzuki, H. Suemastsu, W. Jiang and K. yatsui: Mechanical properties and oxidation behavior of Cr-N-O thin films prepared by pulsed laser deposition, *Transactions of the Materials Research Society of Japan*, **28 (2)**, 421-424 (2003).

[4]H. Asami, J. Inoue, M. Hirai, T. Suzuki, T. Nakayama, H. Suematsu, W. Jiang and K. Niihara: Hardness Optimization of (Cr,Mg)(N,O) thin films prepared by pulsed laser deposition, *Advanced Materials Research*, **11**, 311-314 (2006).

[5]H. Asami, T. Kamekawa, T. Suzuki, T. Nakayama, H. Suematu, W. Jiang and K. Niihara: Preparation of Cr-Cu-N-O thin films by pulsed laser deposition, *Journal of the Japan Insutitute of Metals*, **71(9)**, 704-707 (2007)

[6]J. Shirahata, T. Ohori, H. Asami, T. Suzuki, T. Nakayama, H. Suematsu and K. Niihara: Fourier-transform infrared absorption spectroscopy of chromium nitride thin film, *Japanese Journal of Applied Physics*, **50**, 01BE03 (2011).

Author Index

www.ingramcontent.com/pod-product-compliance
Lightning Source LLC
Chambersburg PA
CBHW072111250125
20788CB00003B/37

9781118205891